EXAM QUESTION PRACTICE PACK

PEARSON EDEXCEL
A-LEVEL GEOGRAPHY

HODDER
EDUCATION
AN HACHETTE UK COMPANY

Orders: please contact Hachette UK Distribution, Hely Hutchinson Centre, Milton Road, Didcot, Oxfordshire, OX11 7HH. Telephone: +44 (0)1235 827827. Email: education@hachette.co.uk. Lines are open from 9 a.m. to 5 p.m., Monday to Friday. You can also order through our website: www.hoddereducation.co.uk.

ISBN: 978 1 5104 7713 1

© Hodder & Stoughton Ltd 2020
First published in 2020 by
Hodder Education
An Hachette UK Company
Carmelite House
50 Victoria Embankment
London EC4Y 0DZ

www.hoddereducation.co.uk

Impression number   10 9 8 7 6 5

Year   2024

Cover photo: mariana_designer - stock.adobe.com

Typeset by Aptara in India

Printed by Hobbs the Printers Ltd, Totton, Hampshire SO40 3WX

A catalogue record for this title is available from the British Library.

www.carbonbalancedprint.com
CBP2250

CONTENTS

INTRODUCTION

This pack of exam-style questions, mark schemes and example responses is specially curated for the Pearson Edexcel A-level Geography specification. The topics covered are:

- Area 1 Dynamic landscapes
  - Topic 1 Tectonic processes and hazards
  - Topic 2 Landscape systems, processes and change: Option 2B Coastal landscapes and change
- Area 2 Dynamic places
  - Topic 3 Globalisation
  - Topic 4 Shaping places: Option 4A Regenerating places
- Area 3 Physical systems and sustainability
  - Topic 5 The water cycle and water insecurity
  - Topic 6 The carbon cycle and energy security
- Area 4 Human systems and geopolitics
  - Topic 7 Superpowers
  - Topic 8 Global development and connections: Option 8A Health, human rights and intervention and Option 8B Migration, identity and sovereignty
- Synoptic investigation
  - Opportunities and challenges for Ethiopia
  - Opportunities and challenges for the Arctic region

The pack is divided into two sections:

- **Exam-style questions**. Questions for each section of each paper are presented in a similar way to the actual exam papers. You may wish to photocopy all or part of them for use with your class.
- **Mark schemes and example responses**. Each question has two student responses – a 'Student A' response typical of an answer that would receive high marks, and a 'Student B' response, which would receive fewer marks. Each response includes a commentary that describes why it receives the marks it does. The mark scheme for each question indicates how the responses could be graded, and can be used alongside each type of student response or just with the question.

This pack is designed to help you to:

- encourage students to reflect on their responses and ensure they know how to succeed
- cultivate students' key skills and knowledge by regular assessment throughout the course or in the revision period before the exams
- incorporate question practice into your lesson plans in the final, vital stage of teaching a topic: putting theory into practice
- allow you to teach flexibly, picking and choosing photocopiable pages as appropriate to share with students
- facilitate peer discussion of what is good or better about given answers, which allows greater insight into quality responses
- allow students to analyse responses without the bias that can come from looking at their own or their friends' work — and so get more from the task

Assessment objectives (AOs)

Assessment objectives are national criteria that are used by examiners when writing exam papers to ensure there is an even coverage of skills. They determine what students are being asked to do and what command terms and key words are being used in the question.

Assessment objective descriptors

Assessment objective	Description
AO1	Demonstrate knowledge and understanding of places, environments, concepts, processes, interactions and change, at a variety of scales
AO2	Apply knowledge and understanding in different contexts to interpret, analyse and evaluate geographical information and issues
AO3	Use a variety of relevant quantitative, qualitative and fieldwork skills to: • investigate geographical questions and issues • interpret, analyse and evaluate data and evidence • construct arguments and draw conclusions

Breakdown of assessment objectives for A-level

	AO1	AO2	AO3	Overall weighting
Paper 1	13%	15.75%	1.25%	30%
Paper 2	13%	15.75%	1.25%	30%
Paper 3	5.5%	6%	8.5%	20%
Non-examination assessment: independent investigation	2.5%	2.5%	15%	20%
Total	34%	40%	26%	100%

EXAM-STYLE QUESTIONS

Area 1 Dynamic landscapes

Resources

Figure 1 Data on tsunami wave run-up heights during the 2011 Japan earthquake and tsunami

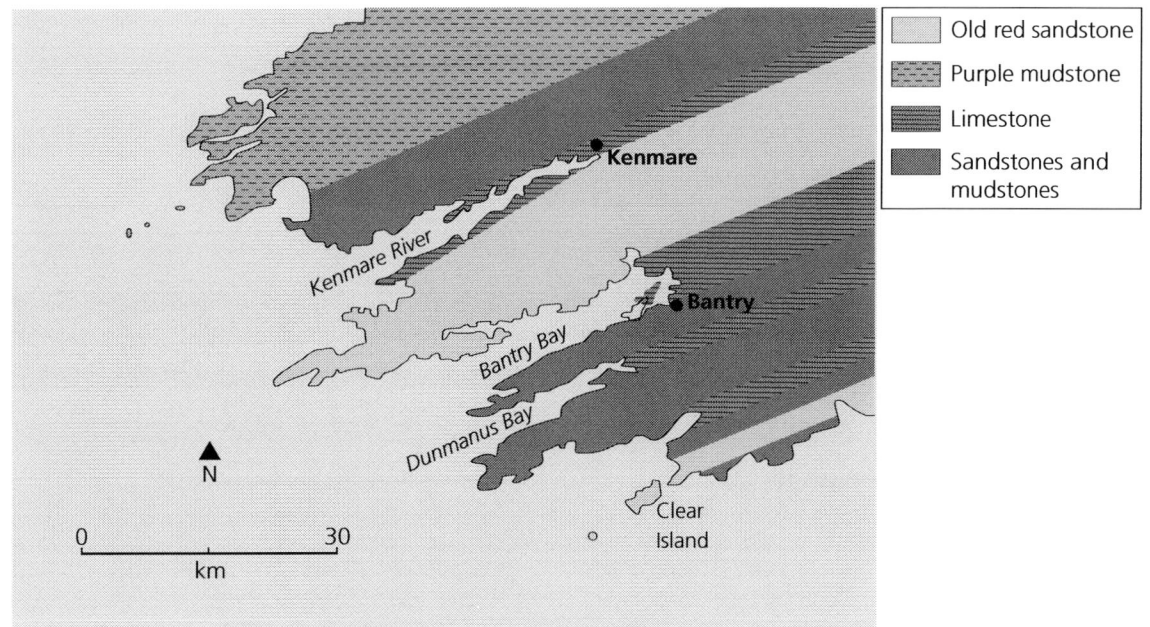

Figure 2 The discordant coast of West Cork, Ireland

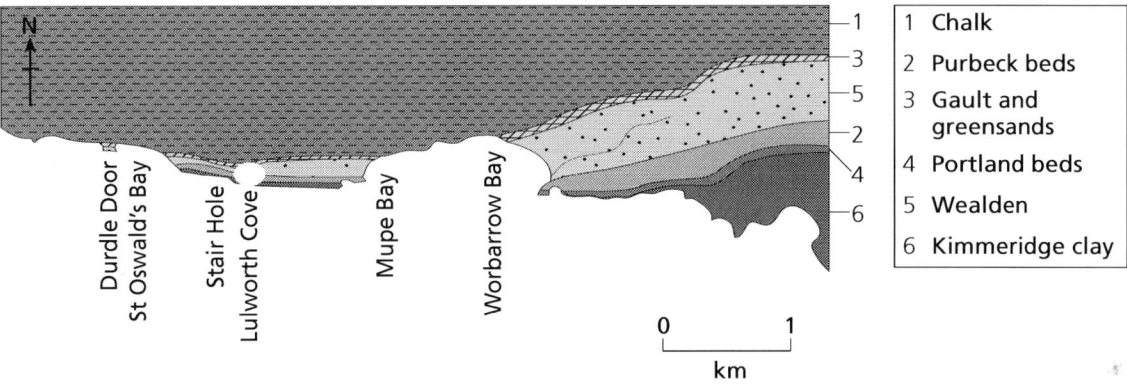

Figure 3 A geological map of the coast around Lulworth Cove, Dorset

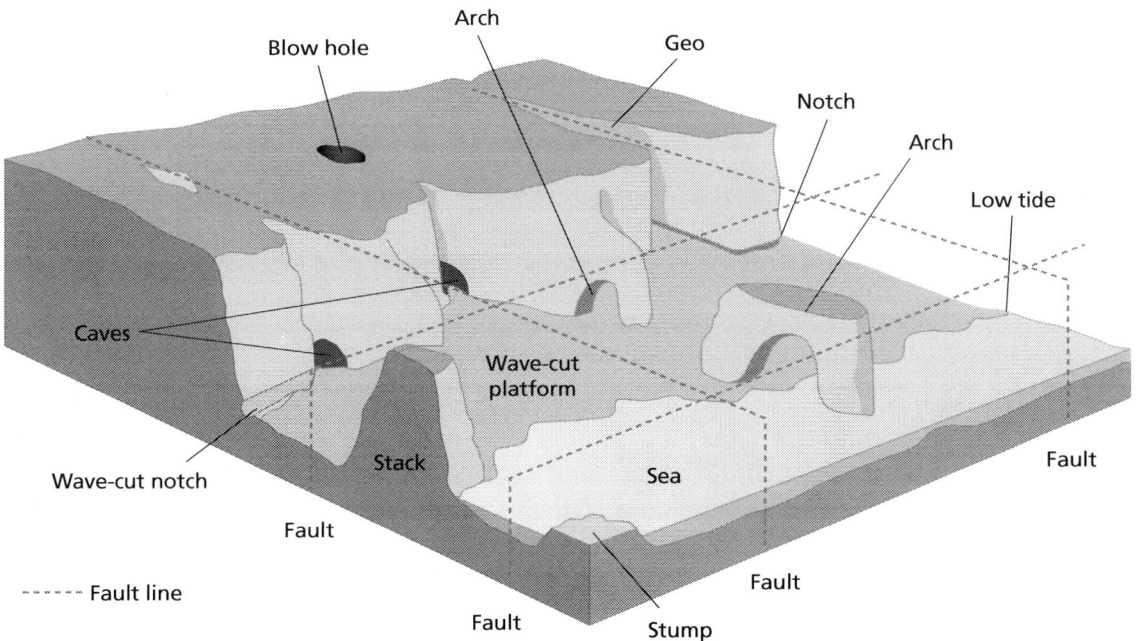

Figure 4 A coastal landscape in resistant sedimentary rocks

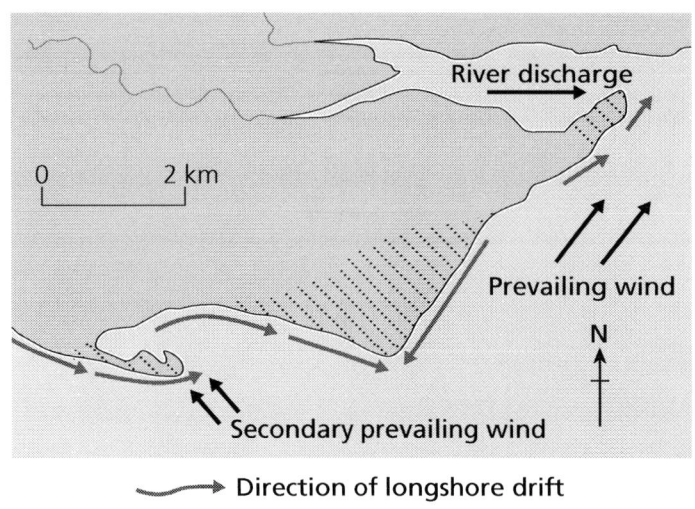

River discharge

0   2 km

Prevailing wind

N

Secondary prevailing wind

→ Direction of longshore drift

Figure 5 Depositional coastal landforms

Topic 1 Tectonic processes and hazards

Questions

1 (a) Study Table 1.

Year	Country	Total number of deaths	Total number of people injured	Total number of people affected
2010	Indonesia	323	455	151,745
2002	DRC	200	400	110,400
2014	Japan	63	69	Unknown
2014	Indonesia	39	72	115,088
2008	Colombia	16	7	22,351
2007	Yemen	6	15	Unknown
2006	Ecuador	5	13	300,250
2007	Ethiopia	5	10	2,000
2011	Indonesia	3	5	2,368
2004	Indonesia	2	105	43,443
2005	El Salvador	2	0	2,000
2005	Comoros (the)	1	0	284,000
2000	Guatemala	0	0	800
2000	Japan	0	0	16,400
2000	Mexico	0	0	41,000

Table 1 Data on 15 volcanic eruptions since 2000

Complete the following table using the data in Table 1.   **(4 marks)**

State the eruption that caused the highest death toll	
Calculate the mean death toll	
State the median value for death toll	
Calculate the percentage of volcanic eruptions that occurred in Indonesia	

(b) Assess the significance of level of development in the effective response to tectonic mega-disasters.

(12 marks)

....

....

....

....

....

....

....

....

....

....

....

....

....

....

....

....

....

....

....

....
....
....
....
....
....
....
....
....
....
....
....
....
....
....

Total: 16 marks

(Mark scheme and example responses on page 101)

2   (a)   Study Figure 1 on page 6. Complete the table below using the data in Figure 1.   **(4 marks)**

State the prefecture which experienced the first tsunami waves	
Calculate the range for the tsunami run-up heights	
Calculate the range for the tsunami run-up heights in Miyagi prefecture	
Calculate the mean tsunami run-up height for Iwate prefecture	

(b) Assess the success of contrasting responses to modify the impacts of tectonic hazards.

(12 marks)

....
....
....
....

Total: 16 marks

(Mark scheme and example responses on page 105)

Topic 2 Landscape systems, processes and change

Option 2B Coastal landscapes and change

Questions

3   (a) (i) Study Figure 2 on page 7. Explain the formation of the discordant coastline shown in Figure 2.   **(6 marks)**

(ii) Study Figure 3 on page 7, which shows the geology of the coast around Lulworth Cove, Dorset. Explain how recession rates may have been affected by alternating rock strata and geological structure along this concordant coastline.   **(6 marks)**

....

....

....

....

....

....

....

(b) Explain how different wave types influence changes in annual beach morphology and sediment profiles. **(8 marks)**

....

....

....

....

....

....

....

....

....

....

....

....

....

(c) Evaluate the relative impacts of short-term coastal flooding and longer-term sea-level change on coastal areas.

(20 marks)

....

....

....

....

Total: 40 marks

(Mark scheme and example responses on page 108)

4 (a) (i) Study Figure 4 on page 7. Explain how erosion processes have contributed to the coastal landscape shown. **(6 marks)**

....

....

....

....

....

....

....

....

....

....

(ii) Study Figure 5 on page 8. Explain the formation of the coastal landforms shown. **(6 marks)**

....

....

....

....

....

....

....

....

....

....

....

(b) Explain the role of vegetation in plant succession on sandy coastlines. **(8 marks)**

....

....

....

....

....

....

....

....

....

....

....

....

....

....

....

....

(c) Evaluate the view that land value is the most important factor influencing policy decisions about how to manage coastlines. **(20 marks)**

....

....

....

....

....

....

Pearson Edexcel A-level Geography Exam Question Practice

Total: 40 marks

(Mark scheme and example responses on page 117)

Area 2 Dynamic places

Resources

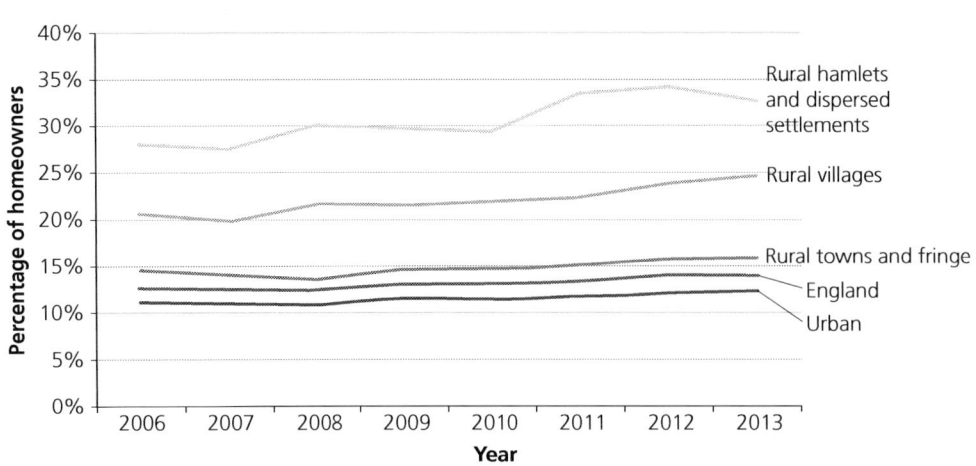

Source: Adapted from the Statistical Digest of Rural England, 2016 (DEFRA). © Crown copyright 2016, reproduced under the Open Government Licence v.3.0

Figure 1 Percentage of homeworkers from all those employed and aged 16 or over, by rural–urban classification, in England, 2006–2013. A homeworker is someone who spends more than half their working time at home

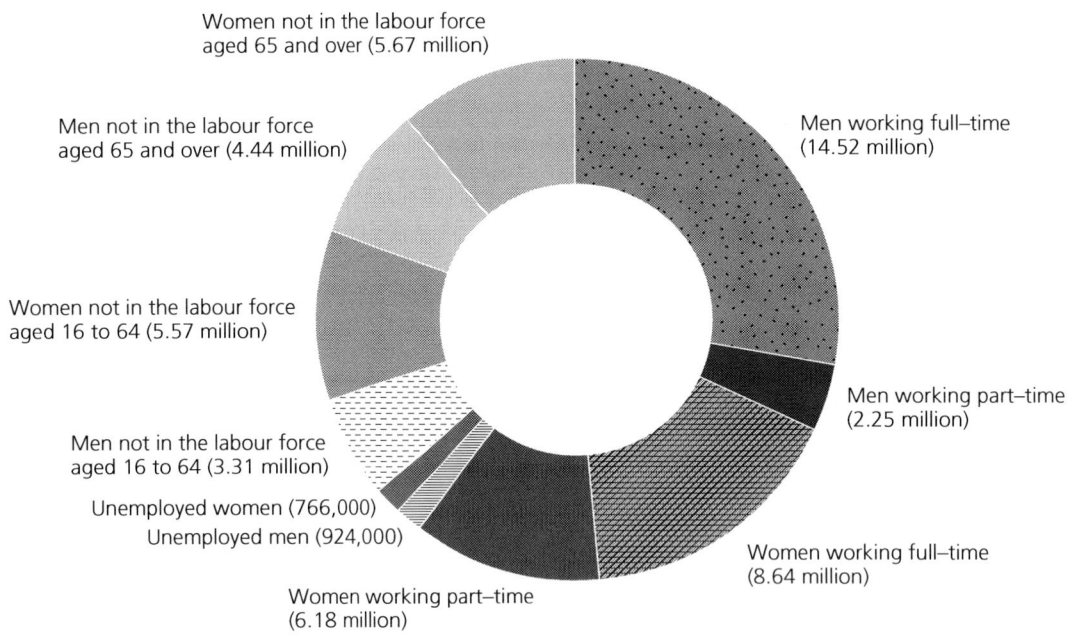

Source: Labour Force Survey — Office for National Statistics, reproduced under the Open Government Licence v.3.0

Figure 2 UK labour market for October to December 2015, seasonally adjusted

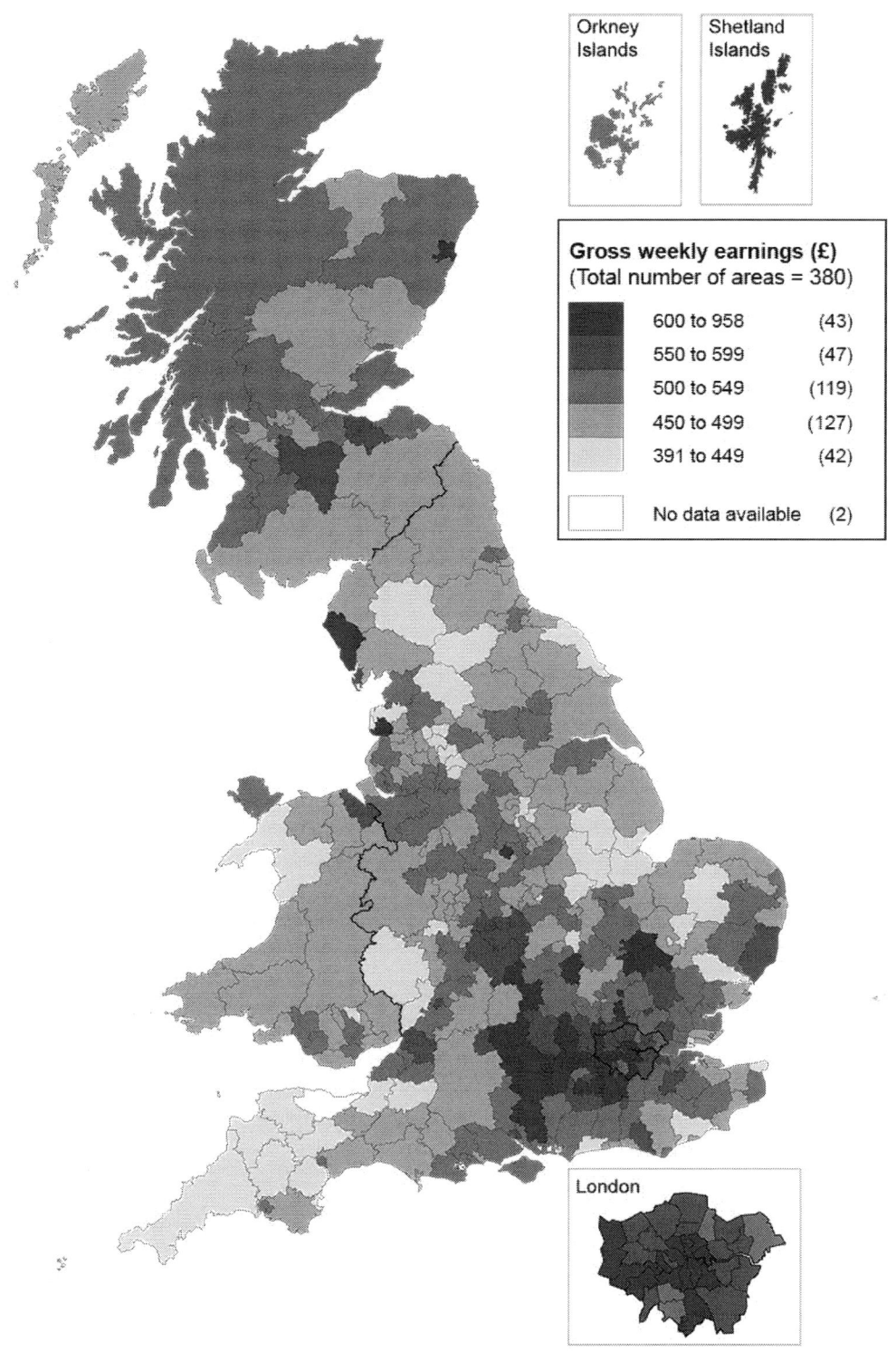

Gross weekly earnings (£)
(Total number of areas = 380)

	600 to 958	(43)
	550 to 599	(47)
	500 to 549	(119)
	450 to 499	(127)
	391 to 449	(42)
	No data available	(2)

Orkney Islands

Shetland Islands

London

Source: Office for National Statistics, Statistical bulletin: Annual Survey of Hours and Earnings: 2016 provisional results, reproduced under the Open Government Licence v.3.0

Figure 3 Median full-time gross weekly earnings by place of work, Great Britain, April 2016

Topic 3 Globalisation

Questions

1   (a)   Explain one reason why globalisation has been accelerated by ICT.   **(4 marks)**

....

....

....

....

....

....

....

     (b) Assess the extent to which the growth of emerging economies has led to both social and environmental challenges for their populations.   **(12 marks)**

....

....

....

....

....

....

....

....

....

....

....

....

....

....

....
....
....
....
....
....
....
....
....
....
....
....
....
....
....
....
....

Total: 16 marks

(Mark scheme and example responses on page 126)

2 (a) Explain one reason why local sourcing by local groups can bring both benefits and costs to the local community. **(4 marks)**

....
....
....

(b) Assess the impacts of international migration on global hub cities. (12 marks)

  Pearson Edexcel A-level Geography Exam Question Practice

....

....

....

....

....

....

....

....

....

....

Total: 16 marks

(Mark scheme and example responses on page 130)

Topic 4 Shaping places

Option 4A Regenerating places

Questions

3   (a) (i) Study Figure 1 on page 20. Suggest one reason for the change in the percentage of homeworkers in rural villages between 2006 and 2013.   **(3 marks)**

....

....

....

....

....

....

(ii) Suggest reasons for the difference in the percentage of homeworkers in urban and rural areas for 2013 shown in Figure 1. **(6 marks)**

....

....

....

....

....

....

....

....

....

....

....

....

(b) Explain why rural places may have changed their demographic characteristics over time. **(6 marks)**

....

....

....

....

....

....

....

....

....

....

....

....

(c) Evaluate the view that the success of strategies used to regenerate urban areas will be viewed differently by different stakeholders.

(20 marks)

 Pearson Edexcel A-level Geography Exam Question Practice

....
....
....
....
....
....

Total: 35 marks

(Mark scheme and example responses on page 133)

4 (a) (i) Study Figure 2 on page 20. Suggest one reason for the difference in the number of men and women working part-time in the UK. **(3 marks)**

....
....
....
....
....
....

(ii) Study Figure 3 on page 21. Suggest reasons why median full-time gross weekly earnings vary across Great Britain. **(6 marks)**

....
....
....
....
....
....
....
....
....

(b) Explain why people's quality of life can be affected by inequalities in pay.　　(6 marks)

(c) Evaluate the view that significant variations in economic and social inequalities can lead to different priorities for regeneration.　　(20 marks)

 Pearson Edexcel A-level Geography Exam Question Practice

Total: 35 marks

(Mark scheme and example responses on page 141)

Area 3 Physical systems and sustainability

Resources

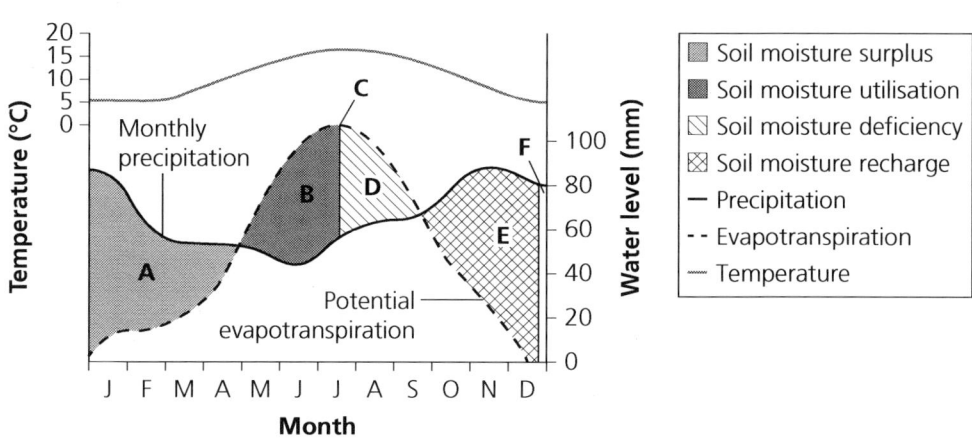

Figure 1 A water budget graph for southern England showing soil moisture status

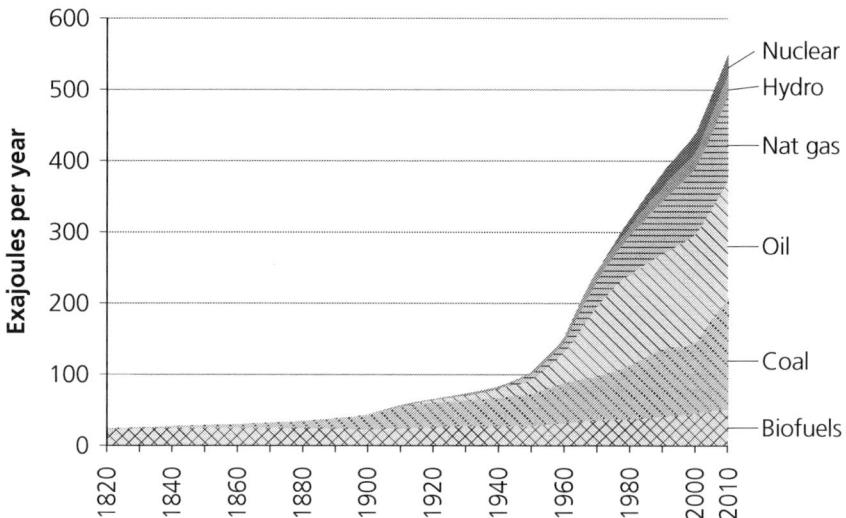

Figure 2 World energy consumption, 1820–2010

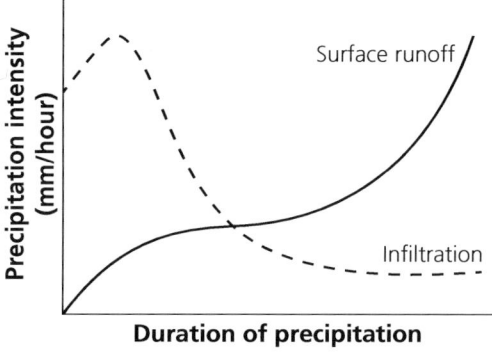

Figure 3 The relationship between precipitation and surface runoff/infiltration

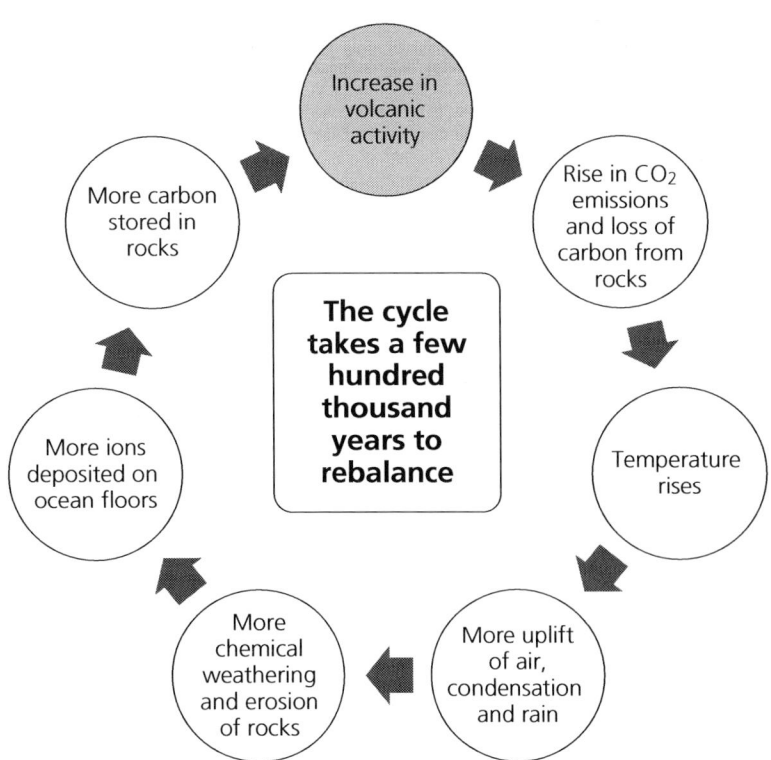

Figure 4 Negative feedback regulating the geological carbon cycle

Topic 5 The water cycle and water insecurity

Topic 6 The carbon cycle and energy security

Questions

1 (a) Study Figure 1 on page 33. Explain the change from soil moisture surplus to soil moisture deficiency in Figure 1.
(3 marks)

....

....

....

....

....

....

(b) Explain how physical factors in a drainage basin affect inputs, flows and outputs. **(6 marks)**

....

....

....

....

....

....

....

....

....

....

(c) Explain how drought is affected by physical and human factors. **(8 marks)**

....

....

....

....

....

....

....

....

....

....

....

....

....

....

....

Area 3 Physical systems and sustainability

(d) Study Figure 2 on page 33. Assess the implications for global energy security of the changing consumption patterns shown.

(12 marks)

....

....

....

....

....

....

....

....

....

....

....

....

....

....

....

....

....

....

....

....

....

....

....

....

(12 marks)

....

....
....
....
....
....
....
....

(e) Evaluate the extent to which climate change caused by human activity is a significant factor threatening the biological carbon and water cycles. **(20 marks)**

....
....
....
....
....
....
....
....
....
....
....
....
....
....

  Pearson Edexcel A-level Geography Exam Question Practice

....

....

....

....

....

....

....

....

....

....

....

Total: 49 marks

(Mark scheme and example responses on page 150)

2　(a)　Study Figure 3 on page 33. Explain the changes in infiltration and surface runoff over time shown on the graph. **(3 marks)**

....

....

....

....

....

....

(b) Explain the physical and human factors that influence the response of a storm hydrograph to a rainstorm. **(6 marks)**

....

....

....

(c) Explain the physical and human factors that can contribute to flooding. (8 marks)

 Pearson Edexcel A-level Geography Exam Question Practice

(d) Study Figure 4 on page 34. Assess the role that volcanic outgassing plays in the long-term geological carbon cycle.

(12 marks)

....

....

....

....

....

....

....

....

....

....

....

....

....

....

....

....

....

....

....

....

....

....

....

....

(e) Evaluate the extent to which alternatives to fossil fuels can help reduce global carbon emissions. **(20 marks)**

Pearson Edexcel A-level Geography Exam Question Practice

....
....
....
....
....
....
....
....
....
....
....
....
....
....
....
....
....

Total: 49 marks

(Mark scheme and example responses on page 161)

Area 4 Human systems and geopolitics

Topic 7 Superpowers

Questions

1   (a)   Explain why IGOs (intergovernmental organisations) are important to global geopolitical stability.   **(4 marks)**

....

....

....

....

....

....

....

(b)   Assess the extent to which the acquisition of physical resources can lead to disputes over ownership and disagreements over exploitation.   **(12 marks)**

....

....

....

....

....

....

....

....

....

....

....

Total: 16 marks

(Mark scheme and example responses on page 172)

2   (a) Explain why emerging countries are increasingly important to global economic and political systems. **(4 marks)**

....

....

....

....

....

....

....

....

(b) Assess the extent to which 'westernisation' has an important influence on the global economic system. **(12 marks)**

....

....

....

....

....

....

....

....

....

....

....

....

....

....

....

....

....

....

....

....

....

....

....

....

....

....

....

....

....

....

....

....

....

....

....

Total: 16 marks

(Mark scheme and example responses on page 176)

     Pearson Edexcel A-level Geography Exam Question Practice

Topic 8 Global development and connections

Option 8A Health, human rights and intervention

Questions

3 **(a) (i)** Study Table 1.

Table 1 Life expectancy in years and the percentage access to improved drinking water for selected countries

Country	Life expectancy (in years)	Access to improved drinking water (%)
Afghanistan	60	55
Australia	83	100
Bolivia	68	90
Botswana	62	96
Chad	51	51
China	75	96
Haiti	62	58
Indonesia	71	87
Venezuela	76	93

Source: World Bank

Using the data from Table 1, complete Figure 1 by plotting the data for Bolivia, China and Indonesia. **(3 marks)**

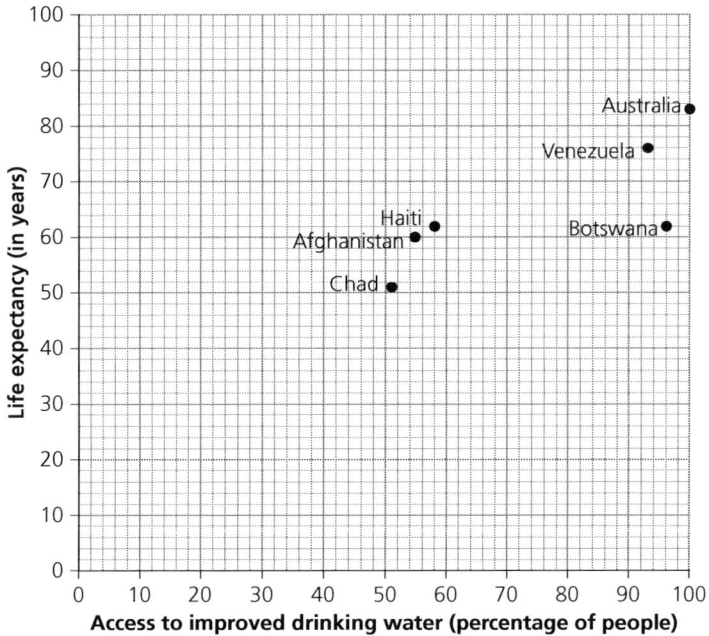

Figure 1 Graph showing the life expectancy in years and the percentage access to improved drinking water for selected countries

(ii) Draw a regression (best fit) line to show the relationship. **(1 mark)**

(b) Suggest reasons for the relationship between life expectancy and percentage access to improved drinking water. **(6 marks)**

....

....

....

....

....

....

....

....

....

....

....

....

....

....

(c) Explain the role of international agreements in promoting human rights. **(8 marks)**

....

....

....

....

....

....

....

....

....

....

....

....

....

 Pearson Edexcel A-level Geography Exam Question Practice

(d) Evaluate the view that development aid leads to problems for developing countries as well as providing them with solutions.   **(20 marks)**

  Pearson Edexcel A-level Geography Exam Question Practice

....
....
....
....
....
....
....
....

Total: 38 marks

(Mark scheme and example responses on page 180)

4   (a) (i) Study Table 2.

Table 2 Selected GDP per capita (PPP) in US$ for Botswana

	2008	**2014**
GDP (gross domestic product) per capita (PPP, purchasing power parity) in US$	13,000	16,000

Source: World Bank

Plot the data from Table 2, showing the US$ GDP per capita (PPP) data for Botswana, on Figure 2. **(2 marks)**

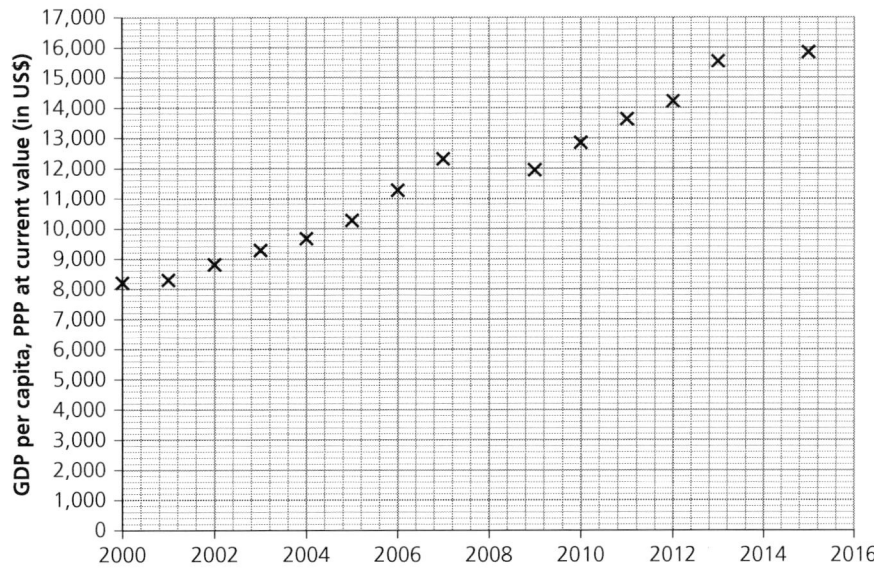

Source: World Bank

Figure 2 Botswana: GDP per capita, PPP (in US$) 2000–2015

(ii) Calculate the percentage growth in GDP per capita (PPP) for Botswana 2000–2015. You must show your working. **(2 marks)**

....

....

....

....

(b) Explain why multiple components are needed to accurately measure human development.

(6 marks)

....

....

....

....

....

....

....

....

....

....

....

....

(c) Explain why levels of political corruption can vary between countries. **(8 marks)**

....

....

....

....

....

....

....

....

....

....

....

....

....

....

(d) Evaluate the view that the UN's Millennium Development Goals have been a globally successful initiative. **(20 marks)**

....

....

....

....

....

....

....

....

....

....

....

....

....

....

....

....

....

  Pearson Edexcel A-level Geography Exam Question Practice

....

....

....

....

....

....

....

....

....

....

....

Total: 38 marks

(Mark scheme and example responses on page 189)

Option 8B Migration, identity and sovereignty

Questions

5 Study Table 3.

Table 3 International migrants as a percentage of the population and the KOF Index of Globalisation for selected countries

Country	International migrants as a percentage of the population	KOF Index of Globalisation (higher = more globalised)
Australia	28	81
Bolivia	1	55
Botswana	7	45
Chad	1	38
China	0.1	60
Haiti	0.2	41
Peru	8.6	65
UK	13.2	82
USA	14	75

Sources of data: World Bank and Axel Dreher, 2006

(a) (i) Using the data from Table 3, complete Figure 3 by plotting the data for the USA, China and Peru.   (3 marks)

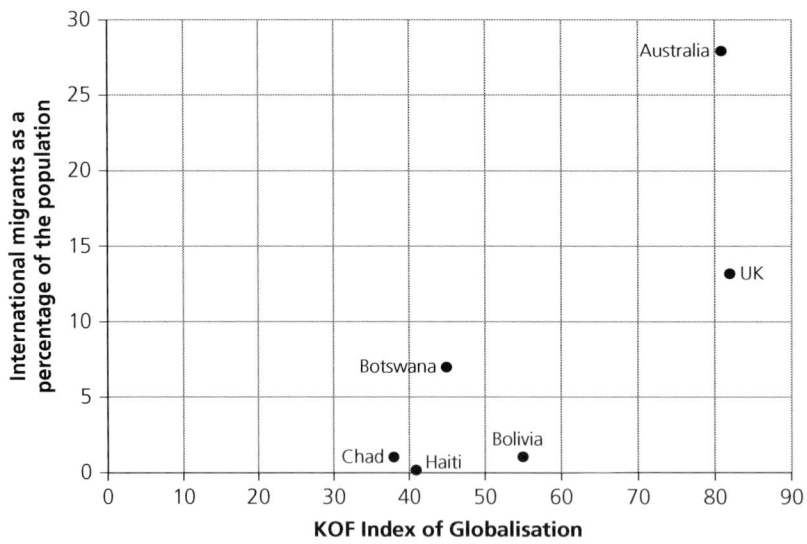

Figure 3 Graph showing international migrants as a percentage of the population and the KOF Index of Globalisation for selected countries

(ii) Draw a regression (best fit) line to show the relationship. **(1 mark)**

(b) Suggest reasons for the relationship between international migrants as a percentage of the population and the KOF Index of Globalisation. **(6 marks)**

....

....

....

....

....

....

....

....

....

....

....

....

(c) Explain why some national borders generate disputes over sovereignty. **(8 marks)**

....

....

....

....

....

....

....

....

....

....

....

....

(d) Evaluate the view that intergovernmental organisations (IGOs) have had limited success when dealing with global environmental problems.

(20 marks)

 Pearson Edexcel A-level Geography Exam Question Practice

....
....
....
....
....
....
....
....
....
....

Total: 38 marks

(Mark scheme and example responses on page 197)

6   Study Figure 4.

Table 4 Selected FDI net earnings abroad (outward) in £million for the UK

2007	2009
92,242	67,942

(a) (i) Plot the data for 2007 and 2009 given in Table 4 showing the UK's foreign direct
investment (FDI) net earnings abroad (outward), onto Figure 4.   **(2 marks)**

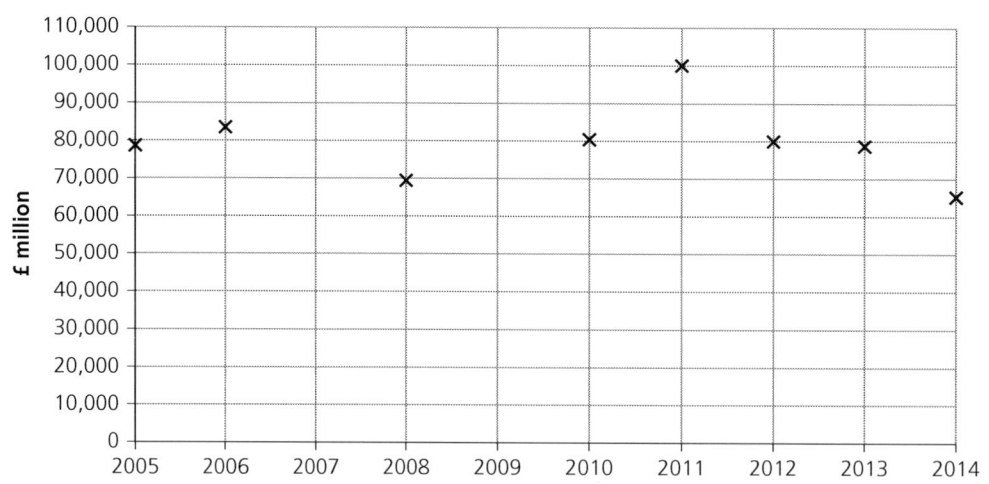

Source: Office for National Statistics, Statistical bulletin: Foreign Direct Investment Involving UK Companies:
2014, reproduced under the Open Government Licence v.3.0

Figure 4 UK's FDI net earnings abroad (outward) in £million, 2005–2014

(ii) Calculate the percentage change in the UK's foreign direct investment (FDI) net earnings abroad (outward) between 2005 and 2014. You must show your working.

(2 marks)

....

....

....

....

(b) Explain why foreign ownership of business and property can have an impact on national identity.

(6 marks)

....

....

....

....

....

....

....

....

....

....

....

(c) Explain why nationalism in the nineteenth century was important in the development of empires.

(8 marks)

....

....

....

....

....

....

(d) Evaluate the view that people with high skill levels are more likely to be able to migrate across international borders. **(20 marks)**

....

....

....

....

....

....

....

....

....

....

....

....

....

....

....

....

....

Total: 38 marks

(Mark scheme and example responses on page 206)

Synoptic investigation

Opportunities and challenges for Ethiopia

Resource Booklet 1

Section A Background information on Ethiopia

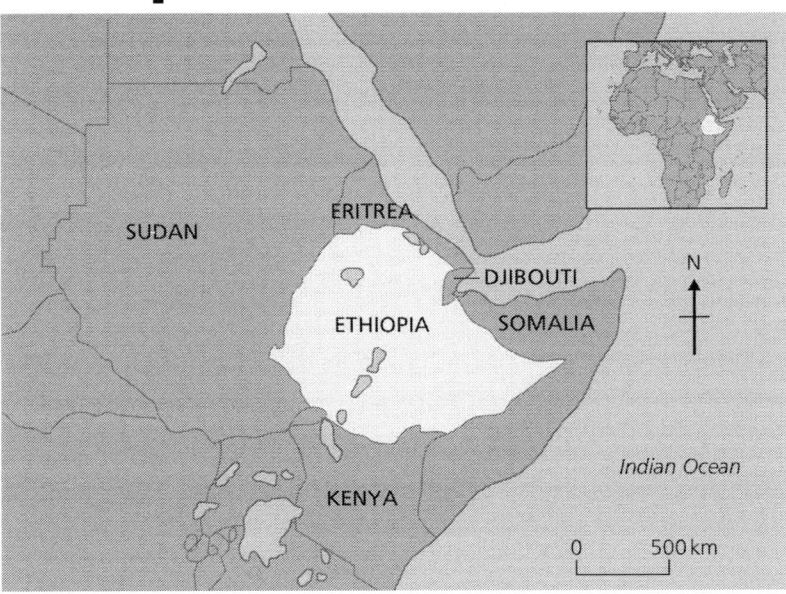

Figure 1 Location map for Ethiopia

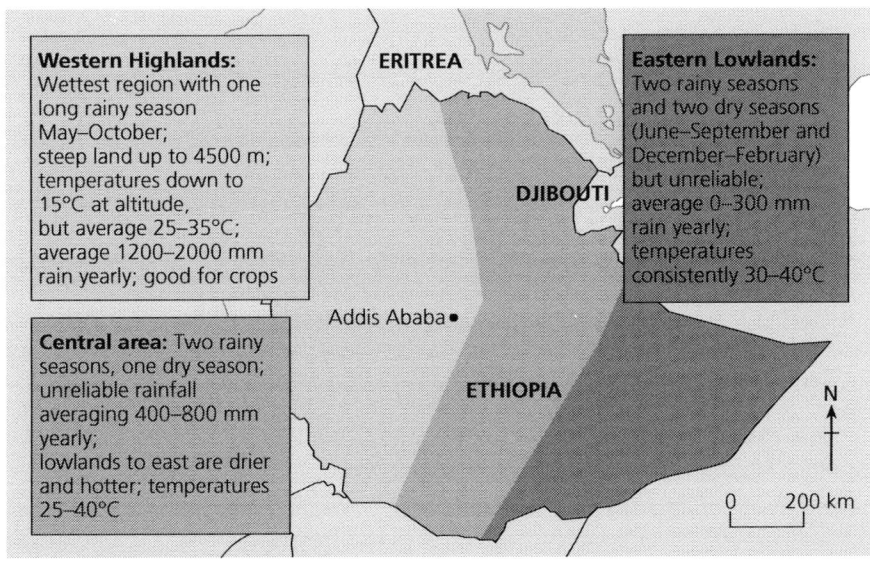

Western Highlands:
Wettest region with one long rainy season May–October;
steep land up to 4500 m;
temperatures down to 15°C at altitude,
but average 25–35°C;
average 1200–2000 mm rain yearly; good for crops

Central area: Two rainy seasons, one dry season;
unreliable rainfall averaging 400–800 mm yearly;
lowlands to east are drier and hotter; temperatures 25–40°C

Eastern Lowlands:
Two rainy seasons and two dry seasons (June–September and December–February) but unreliable;
average 0–300 mm rain yearly;
temperatures consistently 30–40°C

Figure 2 Physical environment of Ethiopia

Date	Key events
Pre-1935	Ethiopia was called Abyssinia and did not come under colonial control, unlike much of Africa.
1935–1941	Italy colonised Ethiopia and invested in transport and power networks.
1941–1974	Rebels and British troops defeated the Italians in 1941 and Ethiopia became independent again. Many years of political instability, conflict, drought and famine. Communism started to grow and a military coup, backed by the Soviet Union and Cuba, took over government.
1974–1987	Over 1.4 million people died in the civil war. Derg government in power and the monarchy abolished. 1977–1978: Ethiopian 'Red terror' – government evicted landowners, migration occurred and economy declined. 50,000 died during Derg government and 1.5 million forced to relocate. 1984–1985: Severe drought and famine led to global action through Live Aid concerts. US$2,000 million given in food aid but Ethiopia still had food insecurity.
1990s	Derg government removed and Ethiopia became more stable, becoming a Federal Democratic Republic. Free trade and tax-free imports of fertilisers and machinery.
2000–present	More support given by USA. Agricultural production has increased and the economy is growing. Five Year Growth and Transformation Plan aimed to reduce poverty. Focus on training farmers with new skills to increase yields. Stable government but some claim that free speech is limited. Increased trust between local authorities and the people, although there have been some demonstrations by some ethnic groups over land and other rights.

Figure 3 Timeline of selected events in Ethiopia

Ethiopia, 2016	
Population	102.3 million
GNI per capita (PPP, US$)	US$590
GDP annual growth rate (%)	6.5%
Life expectancy (years)	62
Birth rate (per 1,000)	37
Death rate (per 1,000)	8
Infant mortality rate (per 1,000 live births)	51
Dependency ratio (dependents aged 0–14 and over 65 per 100 people aged 15–64)	82%
Primary school enrolment (% of primary school-age population)	87%
Secondary school enrolment (% of secondary school-age population)	29%
Adult literacy rate (% ages 15 and over)	49%

   Pearson Edexcel A-level Geography Exam Question Practice

Ethiopia, 2016

Internet users (% of population)	11.6%
Human Development Index	0.442 (174th worldwide)
Motor vehicles (per 1,000)	4

- Growing levels of foreign direct investment with the main investors from the European Union, USA and China.
- Religion: Christianity, 63%; Islam, 33%; traditional, 3%; other, 1% (2007 estimate).
- The capital, Addis Ababa, is sometimes referred to as the 'political capital of Africa'. The city is home to the headquarters of both the Africa Union and the United Nations Economic Commission for Africa as well as other international organisations.

Sources: CIA World Factbook, World Bank Indicators

Figure 4 Fact file for Ethiopia, a low-income developing country (classified by the IMF)

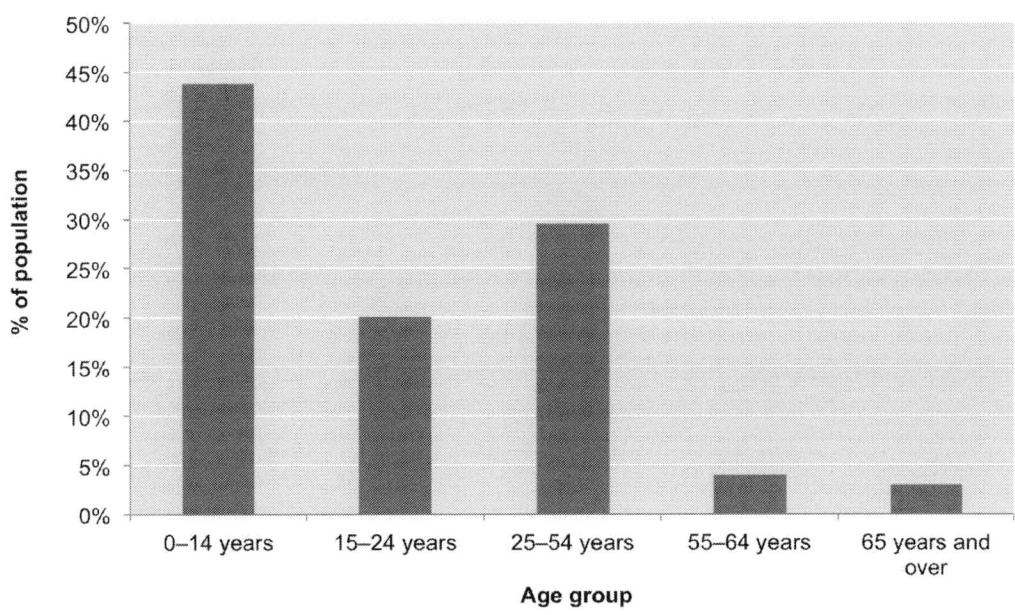

Source: CIA World Factbook

Figure 5 Ethiopia's population by age group (%)

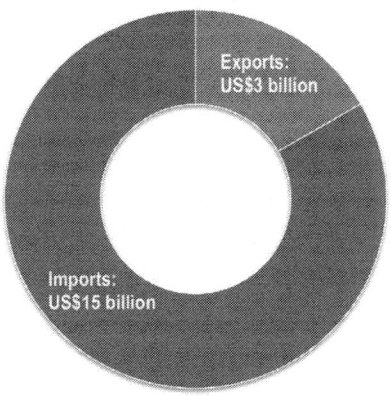

Source: CIA World Factbook

Figure 6 Ethiopia's balance of trade, 2016 (US$billions)

Ethiopia top five imports	Ethiopia top five exports
Metal equipment	Coffee
Electrical materials	Oilseeds
Petroleum oil	Vegetables
Motor vehicles	Gold
Fertilisers	Freshly cut flowers

Source: CIA World Factbook

Figure 7 Ethiopia's top imports and exports, 2015

Section B Comparing the 'Big Five'

Ethiopia can be considered as one of the 'Big Five' countries of Africa, along with Algeria, Egypt, Nigeria and South Africa. These five countries play an important part in regional politics and are likely to significantly influence future developments within the continent because of their population size, their economies and the money that they spend on the military.

Source: Institute for Security Studies, 2015

Country	Algeria	Egypt	Nigeria	South Africa
Population (millions)	39.9	83.4	178.5	53
GNI per capita (PPP, US$)	13,054	10,512	5,341	12,100
Dependency ratio (dependents aged 0–14 and over 65 per 100 people aged 15–64)	42	49	84	45
Internet users (% of population)	18	32	43	49
Infant mortality rate (per 1,000 live births)	21	18	74	32
Life expectancy (years)	74	71	52	57
GDP annual growth rate (%)	3.9	4.2	2.7	3.3

Source: CIA World Factbook

Figure 8 Selected data for Algeria, Egypt, Nigeria and South Africa, 2015

Section C Ethiopia's energy sector

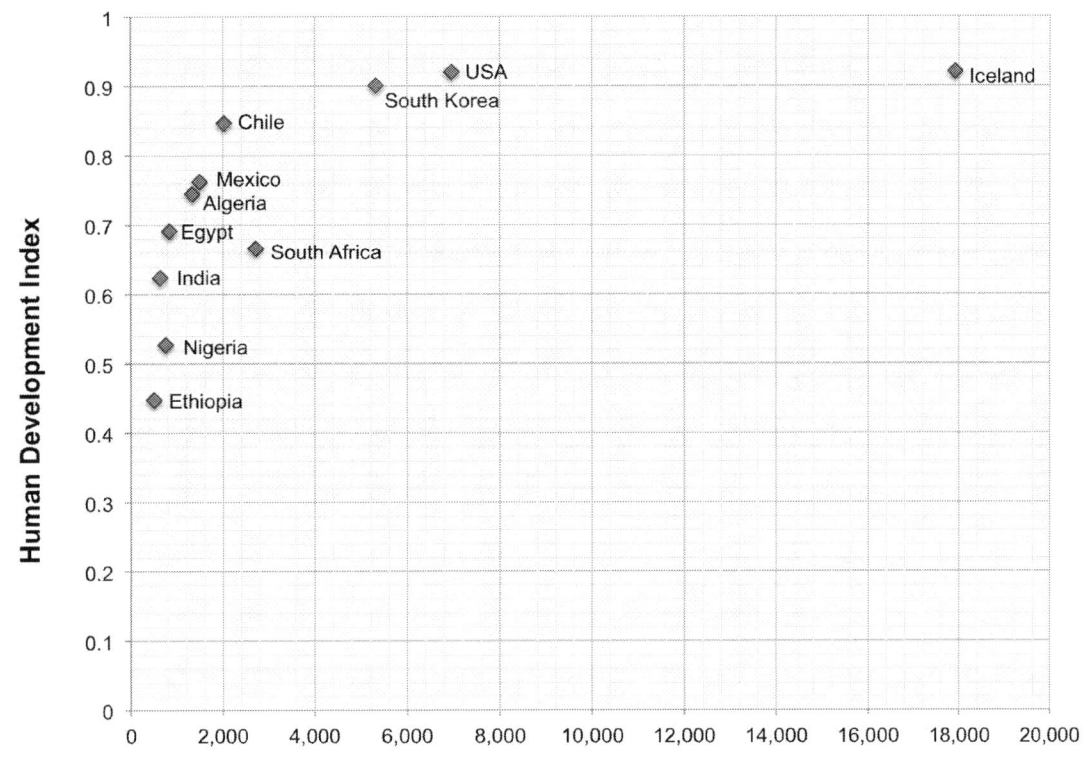

Source: UNDP/World Bank

Figure 9 Human Development Index and energy use per capita, 2015

Source of electricity generation	Percentage
Hydroelectric power	88
Wind power	8
Gas turbines	3.7
Other	0.3

Source: USAid, 2015

Figure 10 Ethiopia's energy mix

> ➤ Existing energy capacity is 2,145 MW. The aim is to increase this to over 8,000 MW.
> ➤ Energy is mainly used in the industrial, commercial and domestic sectors. Agriculture and transport will increase their share during the 2020s.
> ➤ 26% of Ethiopians have access to electricity, with 50% living near the electricity grid.
> ➤ Average electricity consumption by Ethiopians in the home is less than 100 kWh per year.
> ➤ There are around 4 million electricity customers.
> ➤ Biomass for household cooking accounts for 89% of domestic energy consumption.
> ➤ Only 5% of Ethiopia's energy needs are supplied by energy imports.
> ➤ Energy development is part of the Ethiopian government's Growth and Transformation Plan (GTP).

➤ The government's Universal Electrification Access Program (UEAP) aims to connect towns and villages to the electricity grid.

➤ The electricity grid interconnects with parts of Kenya, Sudan and Djibouti, and has the potential for further expansion to Egypt, Tanzania, Rwanda, Burundi and Yemen.

➤ Ethiopia has enormous potential for renewable energy (45,000 MW from HEP; 10,000 MW from geothermal) and aims to become a major energy exporter in the 2030s.

➤ Ethiopia's energy network will need significant infrastructure investment as it grows, e.g. an energy-efficient 'smart grid' system with 132,000 km of distribution lines.

Figure 11 Ethiopia's energy situation

Name	Energy type	Energy capacity	Additional information
Grand Ethiopian Renaissance Dam (GERD)	HEP	6,000 MW	Constructed on the Blue Nile, in northwest Ethiopia. The Nile then flows through Sudan and Egypt. Funded by the Ethiopian government (est. US$5 billion).
Gilgel GIBE III	HEP	1,900 MW	Nearly completed HEP project partly on the Omo river in the southwest of the country. Ethiopian and Chinese funding.
Genale Dawa	HEP	254 MW	Series of dam projects to increase capacity and to increase energy exports to neighbouring Kenya. Funded by Chinese investment.
Adama II	Wind turbine	153 MW	102 turbines located about 90 km southeast of Addis Ababa. Funding through Chinese investment.
Corbetti	Geothermal	1,000 MW	Largest geothermal plant on the continent located in a caldera in the Great Rift Valley. Planned to be completed by 2025. A US$4 billion partnership between Ethiopian government and US/Icelandic consortium.
Reppie	Waste to energy	50 MW	Under construction. First of its kind in Africa. Located in Addis Ababa. Aim is to turn 1,200 tonnes of city waste a day into energy. Part of US President Obama's Power Africa initiative.

Figure 12 Ethiopia's major renewable energy projects

Figure 13 The Blue Nile at Bahir Dar, Ethiopia

Section D Opportunities and challenges
Coffee in Ethiopia

Coffee has been grown in the Oromia region for over 600 years. The Oromia Coffee Farmers' Cooperative Union (OCFCU), established in 1999, represents local cooperatives of farmers based in Oromia. More than 65% of the country's coffee bean-growing land is found in this region. Household farmers sell their coffee beans abroad via the OCFCU. They produce high-quality, organic arabica coffee for export. The coffee can be traced directly back to the grower and is certified by Fair Trade and the Rainforest Alliance. The benefits of the OCFCU can be seen in Figure 14.

Factfile	
Percentage of profits from OCFCU to local cooperatives	70%
Percentage of profits from cooperatives to household farmers	70%
Number of jobs created in local processing factory	2,000
Educational projects financed by OCFCU	• 31 primary schools for 15,000 students • 35 additional classrooms for 6,000 students • 3 kindergartens (nursery schools)
Members' bank created	Financing for pre-harvest and insurance
Growth in numbers of household farmers between 1999 and 2016	22,500 to 33,2000

Figure 14 The Oromia region — the 'birthplace' of coffee

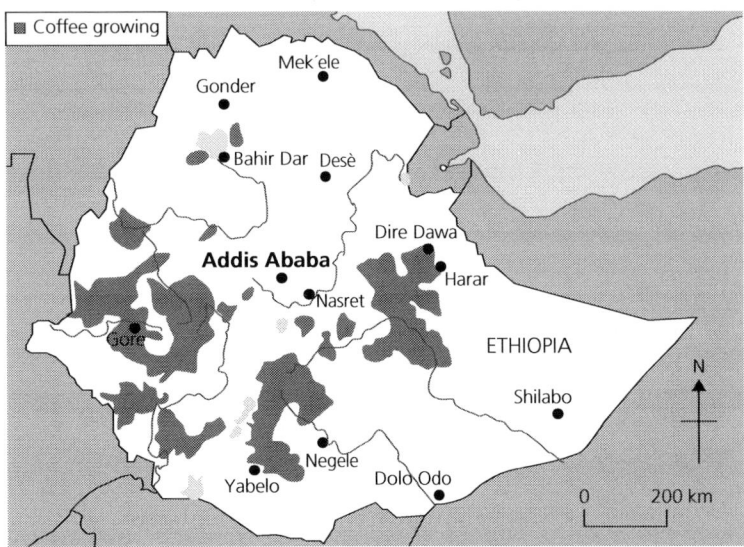

Figure 15 Coffee-growing areas in Ethiopia

Travel, tourism and communication

In the 2010s, Ethiopian Airlines, the national carrier of Ethiopia, expanded to become one of Africa's leading airlines. According to the International Air Transport Association, its profits are higher than all other African airlines combined. While some airlines are experiencing slow growth, Ethiopian Airlines continues to expand its fleet of planes (including 747s and the new Dreamliner) and to increase its number of routes to around 90. It is owned by the Ethiopian government and is able to borrow money cheaply. The airline has benefited from a global drop in the price of fuel as well as having lower labour costs than other airlines. Ethiopian Airlines is also forming partnerships with other regional airlines in Togo and Malawi. Plans are in place for a new four-runway airport in the capital, Addis Ababa. The number of international tourist arrivals increased from 103,000 in 1995 to 770,000 in 2014.

Ethiopian Airlines has also developed a fleet of aircraft for internal flights. Travel times between some areas can be slow by road, so flying is a preferred option for some journeys, particularly for tourists. Chinese investment has led to the upgrading of some of the road network to allow more goods to be transported around the country. There has also been recent investment in the rail network with the reopening of the new electric train between Ethiopia and the port in Djibouti, cutting journey times from 3 days via road to 12 hours.

Figure 16 Ethiopian Airlines internal flight at Lalibela Airport

   Pearson Edexcel A-level Geography Exam Question Practice

Ethiopia has nine UNESCO World Heritage sites, including 11 rock churches that were carved in the thirteenth century at Lalibela. Ethiopia's highlands and plateaux are also becoming increasingly popular for guided trekking.

Figure 17 Church of St George at Lalibela, northern Ethiopia

Figure 18 Refuge for trekkers in Tigray province, northern Ethiopia

Figure 19 Using mobile phones to transfer money in Ethiopia

> Worst drought in decades, with over 10.2 million people needing food aid.
> 9.5% of the population are suffering from food insecurity, with the most vulnerable in the northeast.
> Two consecutive rainy seasons have failed, including the Kiremt rains, which normally feed 80–85% of the country between June and September.
> Livelihoods have been devastated and malnutrition has increased.
> 435,000 children are in need of treatment for severe acute malnutrition.
> 1.7 million children, pregnant women and breastfeeding women are in need of extra food.
> 5.8 million people are in need of emergency water, sanitation and hygiene (WASH) services.
> Water shortages have led to hygiene issues and water-related public health concerns.
> Many areas in the east need additional water to be brought in by trucks.
> School attendance has been affected, with more than 2 million children on the verge of dropping out and over 3,000 schools at risk of closure.
> Children have become more vulnerable.

Source: Reliefweb

Figure 20 The effects of drought in Ethiopia 2015–2016

Question 1 Globalisation

You must use the resource booklet and your own knowledge and understanding from across your course of study to answer this question. Resource Booklet 1 is provided on pages 67–76.

Explain why some locations remain largely 'switched off' from globalisation. **(4 marks)**

....

....

....

....

....

....

....

....

Total: 4 marks

(Mark scheme and example responses on page 215)

Question 2 Spearman's rank correlation coefficient

You must use the resource booklet and your own knowledge and understanding from across your course of study to answer this question. Resource Booklet 1 is provided on pages 67–76.

(a) Table 1 shows data on Gross National Income (GNI) per capita (PPP, US$) and the KOF Index of Globalisation for ten countries.

The formula for Spearman's rank correlation coefficient value R is:

$$R = 1 - \frac{6\Sigma d^2}{n^3 - n}$$

Complete Table 1 and calculate the value of R for the data given. Show your working.

Table 1 Gross National Income per capita and the KOF Index of Globalisation

Country	GNI per capita (PPP, US$)	Rank	KOF Index of Globalisation	Rank	d	d^2
Eritrea	1,130	10	27.13	10	0	0
Ethiopia	1,427	9	37.43	9	0	0
South Africa	12,100	6	64.82	3	3	9
UK	39,200	1	82.96	1	0	0
Algeria	13,054	4	49.36	8		
Egypt	10,512	7	56.33	6	1	1
Nigeria	5,341	8	54.05	7	1	1
Brazil	15,175	3	59.74	5	2	4
China	12,547	5	60.15	4		
Turkey	18,667	2	69.02	2	0	0
					$\Sigma d^2 =$	

Sources of data: UN Development Report (2015) and Axel Dreher (2006)

$R =$ _____ (4 marks)

(b) Explain why the data used to calculate the value of R may not be reliable. **(4 marks)**

....

....

....

....

....

....

....

Total: 8 marks

(Mark scheme and example responses on page 217)

Question 3 Levels of development

You must use the resource booklet and your own knowledge and understanding from across your course of study to answer this question. Resource Booklet 1 is provided on pages 67–76.

Study Figure 8 in Section B of the resource booklet, which shows data for four countries.

Analyse the contrasting development levels of the four countries shown. **(8 marks)**

....

....

....

....

....

....

....

....

....

....

....

....

....

....

....

....

....

Total: 8 marks

(Mark scheme and example responses on page 221)

Question 4 HDI and energy use

You must use the resource booklet and your own knowledge and understanding from across your course of study to answer this question. Resource Booklet 1 is provided on pages 67–76.

Study Figure 9 in Section C of the resource booklet, which shows data on the HDI and energy use per capita for selected countries.

Analyse the relationship between the HDI and energy use per capita.   **(8 marks)**

....

....

....

....

....

....

....

....

....

....

....

....

....

Total: 8 marks

(Mark scheme and example responses on page 223)

Question 5 Ethiopia's energy sector

You must use the resource booklet and your own knowledge and understanding from across your course of study to answer this question. Resource Booklet 1 is provided on pages 67–76.

Study the information in Section C of the resource booklet.

Evaluate the strengths and weaknesses of Ethiopia's energy sector. **(18 marks)**

....

....

....

....

....

....

....

....

....

....

....

....

....

....

....

....

....

....

....

....

....

....

....

....

....

....

....

....

....

....

....

....

....

....

....

....

....

....

....

....

....

....

....

....

....

....

Total: 18 marks

(Mark scheme and example responses on page 225)

Question 6 Ethiopia as a leading economy

You must use the resource booklet and your own knowledge and understanding from across your course of study to answer this question. Resource Booklet 1 is provided on pages 67–76.

Evaluate the view that Ethiopia is in a strong position to become one of Africa's leading economies.

You are advised to use all sections of the resource booklet in your answer to this question. **(24 marks)**

Total: 24 marks

(Mark scheme and example responses on page 229)

 Pearson Edexcel A-level Geography Exam Question Practice

Opportunities and challenges for the Arctic region

Resource Booklet 2

Section A The cryosphere

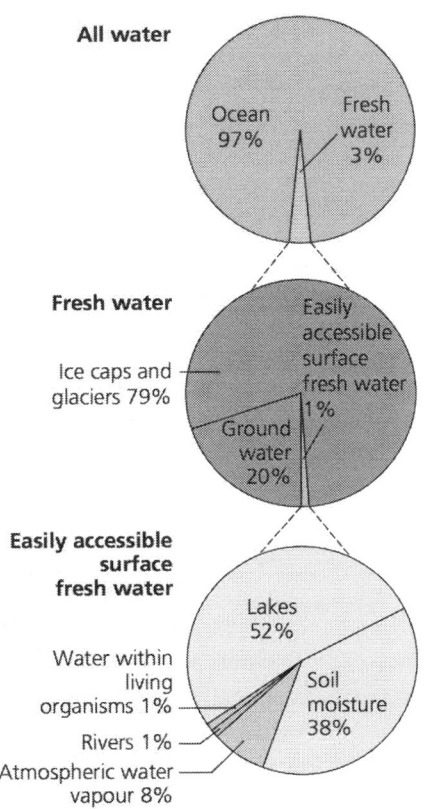

Figure 1 The distribution of the world's water

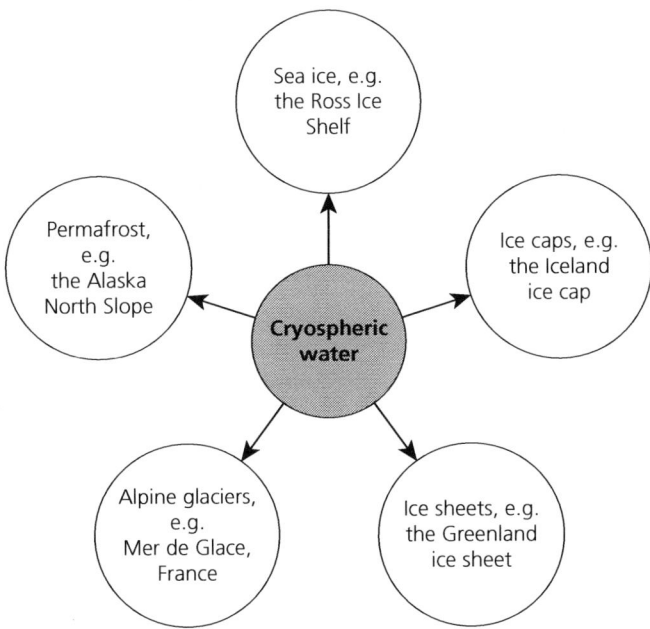

Figure 2 The locations of cryospheric water

Section B Physical environment of the Arctic region

Figure 3 Map of the Arctic

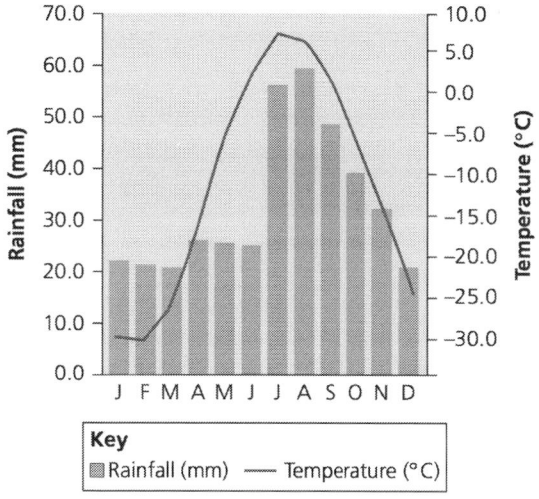

Figure 4 Climate graph for Iqaluit, Northern Canada

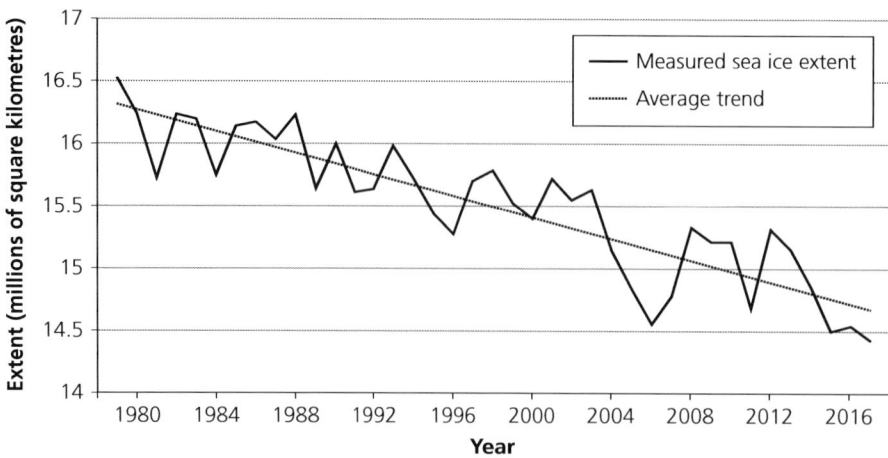

Source: National Snow and Ice Data Center

Figure 5 Average Arctic sea ice extent in November, 1979–2017

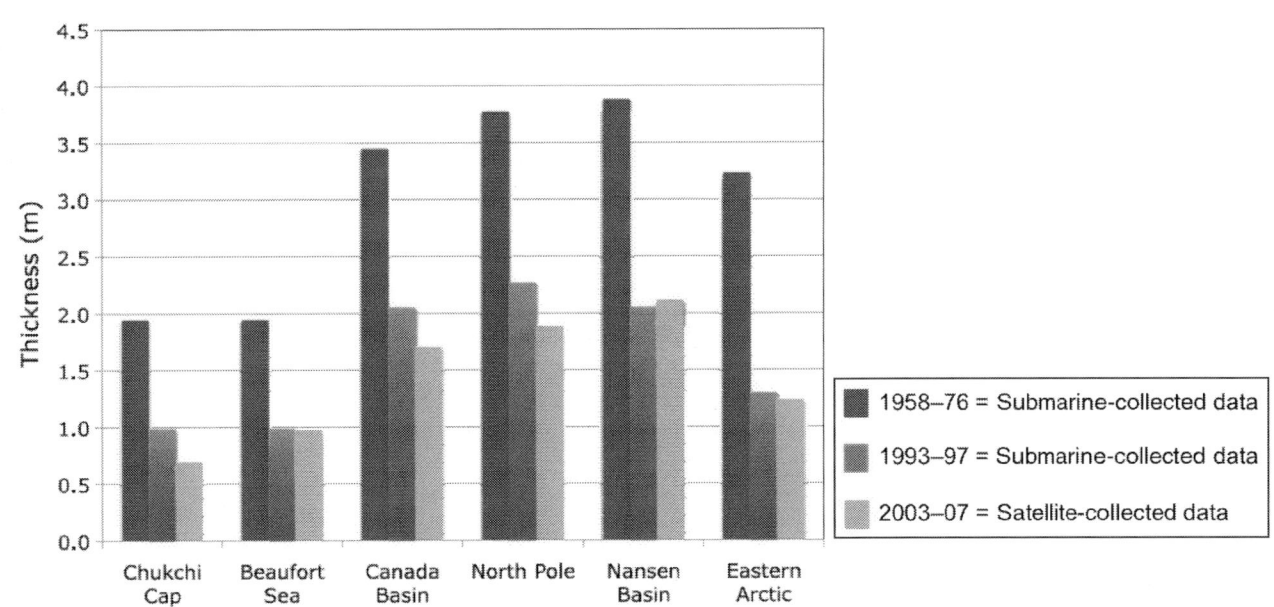

Source: Kwok and Rothrock, 2009. National Snow and Ice Data Center

Figure 6 Changes in thickness of sea ice for selected Arctic areas between the averages for periods 1958–1976, 1993–1997 and 2003–2007

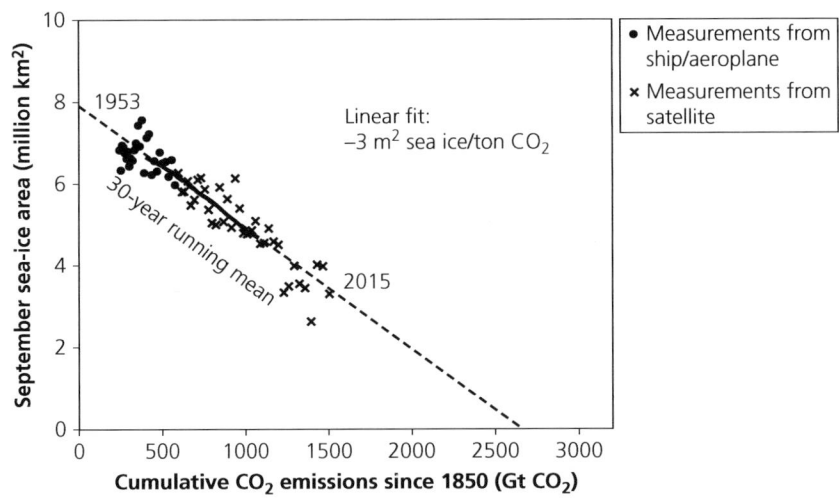

Source: D Notz, Max Planck Institute for Meteorology. National Snow and Ice Data Center

Figure 7 Arctic sea ice and CO_2, 1953–2015

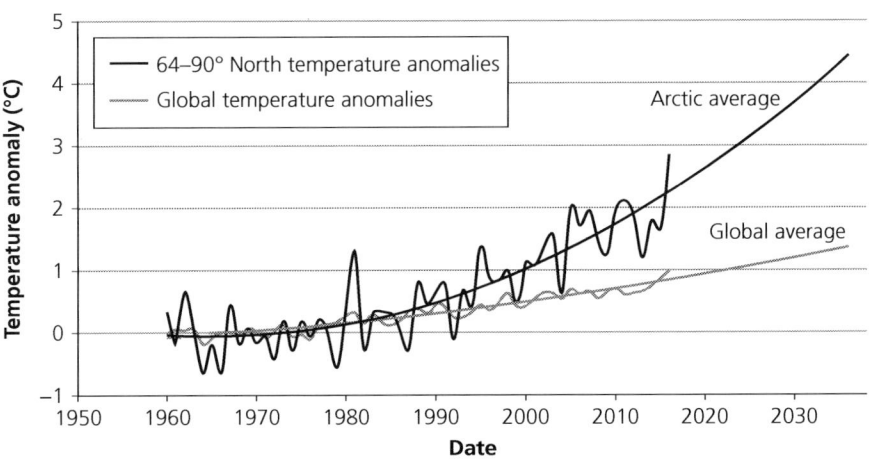

Figure 8 Global versus Arctic temperature anomalies 1960–2017, and projected to 2035

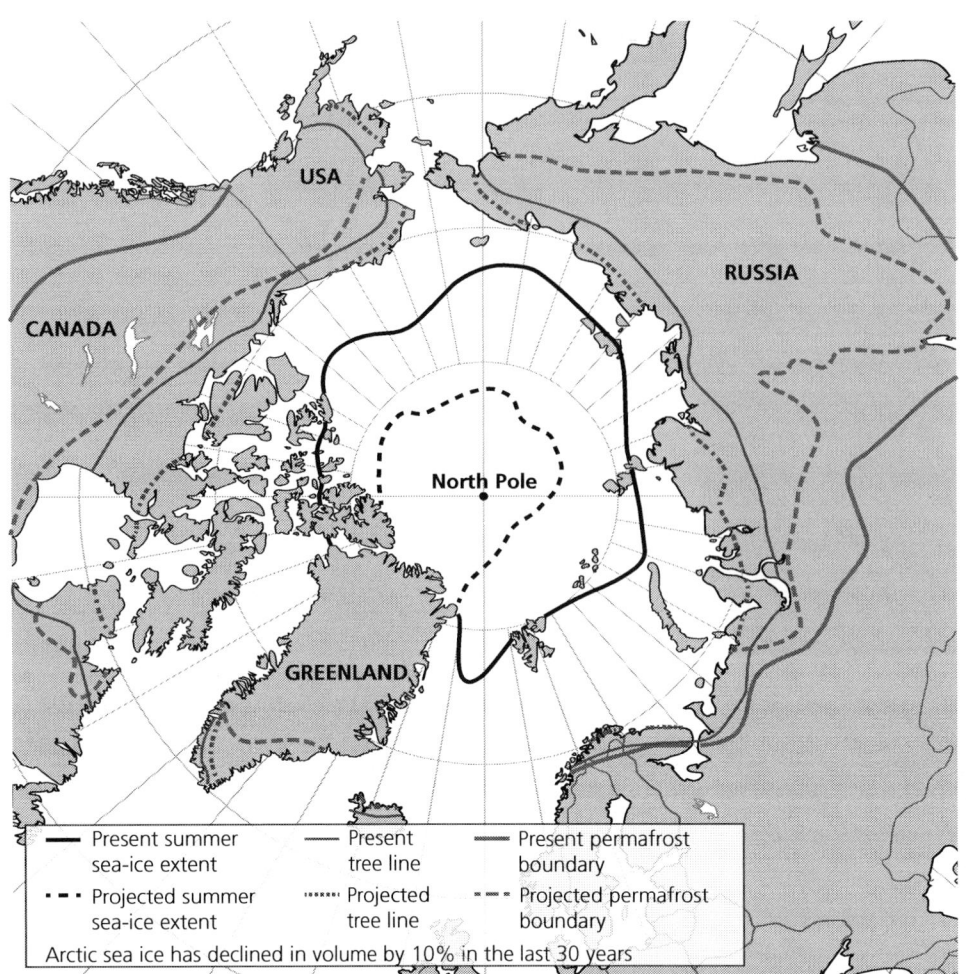

Figure 9 Arctic region: summary of key changes

	1995	2005	2015	2015 R/P*
Total proven reserves oil (thousand million barrels)				
USA	29.8	29.9	55.0	11.9
Canada	48.4	180.0	172.0	107.6
Norway	10.8	9.7	8.0	11.3
Russia	113.6	104.4	102.4	25.5
Natural gas (trillion cubic metres)				
USA	4.7	5.8	10.4	13.6
Canada	1.9	1.6	2.0	12.2
Norway	1.4	2.4	1.9	15.9
Russia	31.1	31.2	32.3	56.3

*2015 R/P is the reserves-to-production ratio for 2015. If the reserves remaining at the end of the year are divided by the production in that year, the result is the length of time that those remaining reserves would last if production were to continue at that rate.

Source: BP Statistical Review of World Energy, 2016

Figure 10 Total proven oil and gas reserves for selected countries, 1995–2015

Section C Selected impacts of the Eyjafjallajökull volcanic eruption in Iceland, April 2016

Figure 11 þorvaldseyri farm at the foot of Eyjafjallajökull volcano, April 2016

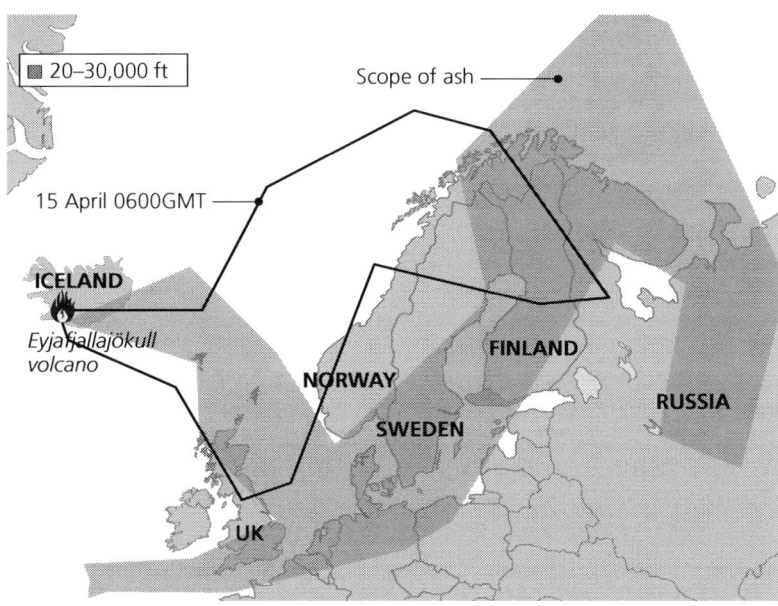

Source: Met Office

Figure 12 Ash plume caused by the eruption of the Eyjafjallajökull volcano, Iceland

Disruptions in air travel

In March 2010, Iceland's Eyjafjallajökull volcano erupted for the first time in over 190 years. By 15 April 2010, the ash plume generated from the eruption had begun to affect much of Europe, spreading as far as northern Italy. The ash cloud grounded flights in most of Europe for several days. More than 100,000 air journeys were cancelled, leading to the worst disruptions in air travel since the 11 September terrorist attack in 2001 (known as 9/11). However, this was a relatively small eruption 'in the wrong place', with no direct deaths. It had a high profile because of the impact on air movements (passenger and freight).

Disruptions in the global supply chain

The 2010 volcano eruption had an impact on the global supply chain. Imports and exports were greatly affected by the air travel shutdown. Although airfreight accounts for a tiny amount of world trade by weight, it accounts for a much higher proportion of trade by value. For example, airfreight accounts for approximately 0.5% of UK trade by weight but 25% of trade by value.

Disruptions in car manufacturing

The disruption to airfreight by the eruption highlighted how important airfreight is in supplying high-value key components to many manufacturers. The Nissan plant in Japan, for example, had to stop production of the Cube, Murano and Rogue crossover models because it ran out of a critical sensor that is produced in Ireland. Airfreight is only used for a small quantity of high-value but vital electronic components where there are few alternative suppliers.

Disruptions in the transportation of perishable goods

There were impacts on the producers of flowers, fruit and vegetables in African countries such as Kenya, Zambia and Ghana, with delays in transportation meaning large quantities of fast-perishing produce rotted, resulting in a loss for producers. The World Bank estimated that, in total, African countries may have lost US$65 million because of the effect of the airspace shutdown on perishable exports.

Figure 13 Selected economic impacts of the Eyjafjallajökull volcanic eruption, April 2010

> The Eyjafjallajökull volcano erupted in 2010, creating a large ash cloud across much of Europe and the North Atlantic. This had a significant impact on air travel as ash can be a serious danger to aircraft and can lead to engine failure.

> The London Volcanic Ash Advisory Centre (VAAC), part of a network of nine VAACs worldwide set up by the International Civil Aviation Organization (ICAO), issued advice from specialist forecasters about the current and predicted location of the ash from the eruption. These data were gained from sources such as satellite, aircraft and ground observations as well as weather forecasting and ash dispersion models.

> The London VAAC identified the ongoing risk to aircraft and it was predicted that the ash would be present over much of Europe and the North Atlantic. In over 24 countries 300 airports were closed, as was much of Europe's airspace between 15 and 21 April. This had a huge impact on worldwide air travel. 7 million people were affected as more than 100,000 flights were cancelled, and it is estimated that airlines lost US$1.7 billion in revenue.

> New advice was issued so that different concentrations of ash could be zoned. Aircraft could be flown in areas with low concentrations of ash (<2–4 mg/m³) with minimal risk, providing airlines conducted more risk assessments and aircraft inspections. However, as the scientific validity of these ash concentration charts has been questioned, they are only in use within Europe.

Figure 14 Impact of the Eyjafjallajökull volcanic eruption on European airspace, 2010

The eruption had a significant effect on Icelandic tourism in 2010, with the number of Norwegians, Danes, Germans and Swedes decreasing from previous years.

Figure 15 Impact on Icelandic tourism, 2010

Number of passengers carried by air transport: **2.628 billion**

Number of air transport departures worldwide: **29.6 million**

Figure 16 Global air transport figures, 2010

Section D Selected social, economic and political data on the Arctic region

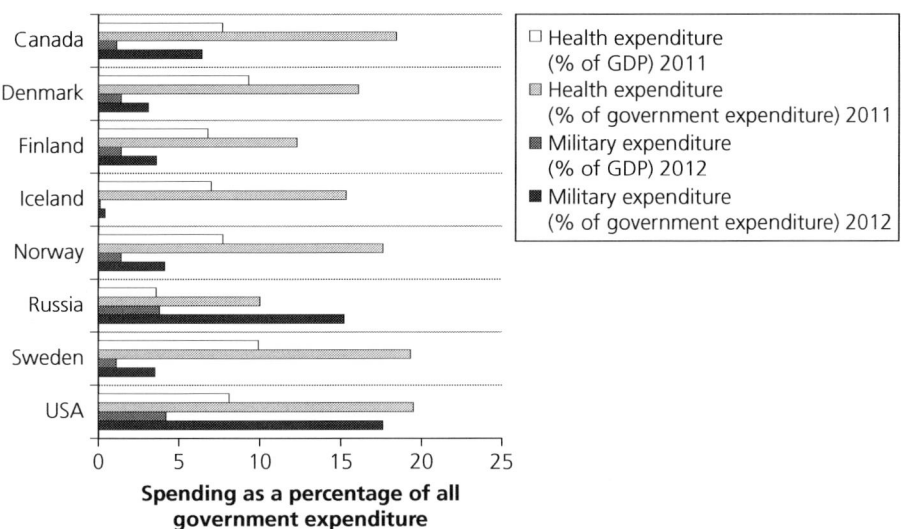

Source: World Bank

Figure 17 Military and health expenditure as a percentage of government expenditure and as a percentage of GDP for selected countries, 2011–2012

Many indigenous groups, such as Inuit and Sami, have lived in the Arctic for thousands of years. Some groups are nomadic while others hunt for marine animals. They are resilient and able to survive in extreme conditions by adapting their diet and making use of the resources. Some indigenous groups had land and resources that 'belonged' to them taken by force. During the Cold War, military resources were installed in the area. Today, the interests of indigenous people are represented by organisations such as the Inuit Circumpolar Council.

Figure 18 Indigenous people

➤ The Arctic is strategically important. Several countries have 'claimed' areas of the Arctic based on their geographical location, geology or historical claims.

➤ Transportation routes close to both Greenland and the Russian coast could open up with continued melting of the sea ice. This would make journeys from Europe to Asia up to 40% faster and would avoid having to travel through some Middle Eastern routes that may be more politically unstable and susceptible to piracy.

➤ The Arctic contains much of the world's unexploited oil reserves (13% of oil and 30% of gas). Exploration could become more economically viable with increased ice melting and improved technology.

➤ Changes in the Arctic climate could open up the region for tourism, particularly cruising and fishing.

➤ The fragile Arctic ecosystem and its species could be affected by climate change.

Figure 19 Selected Arctic issues

The Arctic Council is an intergovernmental organisation aimed at promoting cooperation among its members. Below is a summary of some of the main interests of each member.

Country	Selected priorities in the Arctic
Canada	• Climate change • Relations with indigenous people • Management of the Arctic Ocean • Sustainable economic and social development
Kingdom of Denmark	• Peaceful, secure and safe Arctic • Self-sustaining growth and development • Development with respect for the Arctic's vulnerable climate, environment and nature • Close cooperation with international partners
Finland	• Interdisciplinary research on global changes and their consequences for the Arctic environment • Arctic technology, e.g. construction and transport • Improving the status of the indigenous population (Sami)
Iceland	• Securing Iceland's position within the Arctic region • Resolving differences, e.g. fisheries, navigation, mineral exploitation • Increasing cooperation with other states • Furthering security, trade and knowledge
Norway	• Research on climate and the environment • Diversified economy, e.g. fish farming, liquefied natural gas processing, tourism, minerals • Supporting the rights of the Sami • Working to make the Arctic a peaceful region of cooperation and sustainable resource management
Russia	• Efficient and sustainable development of the Arctic • Using the Arctic Zone as a strategic resource base • Creating an area of peace and cooperation • Using the North Sea route and improving information infrastructure • Socio-economic development for the indigenous population
Sweden	• Promotion of economically, socially and environmentally sustainable development • Supporting the culture and rights of the Sami population of 20,000 • Climate and environmental research • Ice breaking to support commercial shipping and environmental monitoring
USA	• National and homeland security • Cooperation with other Arctic states • Involvement of the indigenous Alaskans • Supporting scientific research

Source: www.arctic-council.org

Figure 20 Governance of the Arctic region

 Pearson Edexcel A-level Geography Exam Question Practice

Question 1 The cryosphere as a water store

You must use the resource booklet and your own knowledge and understanding from across your course of study to answer this question. Resource Booklet 2 is provided on pages 85–92.

Explain why the cryosphere is an important water store. **(4 marks)**

....

....

....

....

....

....

....

Total: 4 marks

(Mark scheme and example responses on page 233)

Question 2 Changes in Arctic sea ice

You must use the resource booklet and your own knowledge and understanding from across your course of study to answer this question. Resource Booklet 2 is provided on pages 85–92.

(a) Figure 5 and Figure 6 in the resource booklet show data relating to changes in Arctic sea ice.

    (i) Using the data in Figure 5, calculate the percentage change in Arctic sea ice extent between 1979 and 2015. **(2 marks)**

....

....

....

    (ii) Calculate the difference in metres in the average sea ice thickness between 1958–1976 and 2003–2007 for the North Pole, as shown in Figure 6.

    You must show your working. **(2 marks)**

....

....

....

....

(b) Explain why the data used to calculate the values of sea ice extent and thickness may be unreliable.

(4 marks)

....

....

....

....

....

....

....

Total: 8 marks

(Mark scheme and example responses on page 235)

Question 3 Reserves in oil and gas

You must use the resource booklet and your own knowledge and understanding from across your course of study to answer this question. Resource Booklet 2 is provided on pages 85–92.

Study Figure 10 in the resource booklet showing total proved oil and gas reserves for selected countries.

Analyse the differences in reserves for both types of energy source for the USA, Canada, Norway and Russia.

(8 marks)

....

....

....

....

....

....

....

....

....

....

....

....

....

....

....

Total: 8 marks

(Mark scheme and example responses on page 237)

Question 4 Expenditure on military and health

You must use the resource booklet and your own knowledge and understanding from across your course of study to answer this question. Resource Booklet 2 is provided on pages 85–92.

Study Figure 17 in Section D of the resource booklet, which shows data on military and health expenditure as a percentage of government expenditure for selected countries.

Analyse the data shown in the graph.   **(8 marks)**

....

....

....

....

....

....

....

....

....

....

....

....

Total: 8 marks

(Mark scheme and example responses on page 239)

Question 5 The Eyjafjallajökull eruption

You must use the resource booklet and your own knowledge and understanding from across your course of study to answer this question. Resource Booklet 2 is provided on pages 85–92.

Study the resources in Section C of the resource booklet.

Evaluate the global impact of the eruption of the Eyjafjallajökull volcano in April 2010.   **(18 marks)**

....

....

....

....

....

Total: 18 marks

(Mark scheme and example responses on page 242)

Question 6 Future opportunities and challenges

You must use the resource booklet and your own knowledge and understanding from across your course of study to answer this question. Resource Booklet 2 is provided on pages 85–92.

Evaluate the view that the Arctic region faces both opportunities and challenges in the future.

You are advised to use all sections of the resource booklet in your answer to this question.

(24 marks)

....

....

....

....

....

....

....

....

....

....

....

....

....

Total: 24 marks

(Mark scheme and example responses on page 245)

MARK SCHEMES AND EXAMPLE RESPONSES

Area 1 Dynamic landscapes

Topic 1 Tectonic processes and hazards

Question 1 mark scheme

(a) 4 marks (AO3 = 4 marks)
- ➢ State the eruption that caused the highest death toll = 2006 Ecuador eruption
- ➢ Calculate the mean death toll = 44.3
- ➢ State the median value for death toll = 5
- ➢ Calculate the percentage of volcanic eruptions that occurred in Indonesia = 26.7

Hints and tips
When calculating, give your answer to 1 decimal place.

(b) 12 marks (AO1 = 3 marks, AO2 = 9 marks)
Some suggested ideas are given below but you may wish to expand on these or include other relevant points.

AO1 Demonstrating your knowledge and understanding
- ➢ Mega-disasters produce large-scale impacts, which have regional or even global consequences.
- ➢ The effect they have on people is determined to a large extent by the level of development of the country/countries involved.
- ➢ There are other factors that are significant in terms of effective response, such as the magnitude of the event, the hazard profile of the event, and rescue and aid efforts.

AO2 Applying your knowledge and understanding
- ➢ Make relevant connections.
- ➢ Support your evaluation with evidence.
- ➢ Produce a balanced and coherent argument.

Development can negatively affect the vulnerability of people leading up to the hazard. For example:
- ➢ Unsustainable development increases vulnerability.
- ➢ Lack of money can result in under-investment in resilience training for the population.
- ➢ Issues with social development, e.g. poor health/nutrition levels.

Development can also negatively impact the post-disaster situation:

➤ Damage to infrastructure such as homes and services
➤ Social — damage to provision of health and education

Development can also influence preparedness in a positive way, for example:

➤ Investment in hazard monitoring
➤ Infrastructure resilience
➤ Mitigation strategies, e.g. tsunami walls
➤ Training and preparedness of local population

Following the disaster there are opportunities for development to positively influence future resilience through, for example, rebuilding strategies that make buildings more resilient.

Other factors can have a significant impact on how effective the response is, such as:

➤ Prediction: Volcanic eruptions are often predicted, leading to evacuation, whereas earthquake mega-disasters cannot be predicted. The 2004 Asian tsunami mega-disaster was not predicted, and while warnings were issued for the 2011 Sendai tsunami, impacts were still enormous.
➤ Mega-disaster magnitude events can overwhelm even well prepared, high-income countries, e.g. the 2011 Japanese earthquake and tsunami.
➤ Rapid rescue, relief and aid response can reduce some of the long-term impacts; international aid from governments and NGOs could offset lack of preparation and response in some developing countries, up to a point.

Answers to this question will be given a mark within a level band

Level 1 (1–4 marks) You show only a limited geographical knowledge and understanding of the significance of development. You make limited connections between aspects of your answer and support your interpretations with limited evidence. You draw unbalanced conclusions based on the material in your answer.

Level 2 (5–8 marks) You show mostly relevant and accurate geographical knowledge and understanding of the significance of development. You make mostly relevant connections between aspects of your answer as appropriate and support your interpretations with some evidence. You draw conclusions based on the material in your answer but your conclusions may be limited or unbalanced.

Level 3 (9–12 marks) You show relevant and accurate geographical knowledge and understanding of the significance of development. You make sound connections between aspects of your answer as appropriate and support your interpretations logically with evidence. You draw balanced and logical conclusions based on the material in your answer.

Hints and tips

Assess the ways in which development affects the effectiveness of the response to tectonic mega-disasters. You should review various pieces of information and draw them together in a conclusion at the end.

Question 1 example responses

Student A

(a)

State the eruption that caused the highest death toll	2006 in Ecuador
Calculate the mean death toll	44.3
State the median value for death toll	5
Calculate the percentage of volcanic eruptions that occurred in Indonesia	26.7

ⓔ **All four answers are correct.** **4 marks**

(b) Mega-disasters cause large-scale impacts, in terms of either the area they cover or the extent of the impacts on the people involved. These impacts frequently affect more than one country and have regional or even global significance. In developing and emerging countries, lower levels of development can have a large effect on how significant the impacts of a hazard are. But where development levels are higher, the impacts of hazards can be reduced.

Lower levels of development can negatively affect the vulnerability of the local populations. For example, unsustainable development can increase vulnerability of local people. In Haiti prior to the 2010 earthquake, rapid rates of urbanisation had led to the growth of shanty towns in Port au Prince. These buildings were less able to withstand the shaking of the earthquake and many collapsed. Even in the longer-established buildings, lack of proper design and poor construction techniques resulting from poverty caused many of the buildings to pancake as the floors collapsed on top of each other. In addition, lack of money can result in under-investment in preparing the population to deal with hazards when they occur. Furthermore, low levels of social development can make a population more vulnerable. If nutrition and health levels are lower to begin with, following an earthquake the population will be more vulnerable to diseases that readily spread in the makeshift camps people set up. For example, in Haiti in 2010, an additional 9,000 people died as a result of a cholera outbreak following the earthquake.

Development can affect the response to the threat of hazards in a positive way too. For example, in richer countries money can be invested in monitoring hazards. In Japan before the 2011 earthquake, a network of 182 seismometers constantly monitored the country for signs of seismic activity. Again in Japan, strict building codes ensured that very few buildings collapsed due to seismic shaking during the 2011 magnitude 9.0 earthquake. Additionally, mitigation strategies can be implemented. The Japanese have an extensive sea-wall defence against the tsunami threat, however there is evidence that these walls led to complacency among the local population during the 2011 quake, resulting in many thousands of deaths. In addition, the walls were not high enough to stop the 2011 tsunami so economic losses were the largest in human history from a tectonic disaster. The 2010 eruption of Eyjafjallajökull in Iceland caused widespread economic losses due to flight disruption across Europe and North America caused by an extensive ash cloud that grounded aircraft. Most countries affected were developed world ones, but there was no effective management strategy despite the wealth and technology of the countries affected.

The nature of the tectonic event is also important. Volcanic eruptions can usually be predicted, so warning and evacuation is possible. This is, however, not the case with earthquakes, which occur without warning. This means longer-term preparations are all the more important, such as education, building codes, rescue and recovery preparations — and unfortunately these are generally less widespread in lower-income countries.

Overall, then, development can have a significant effect, either positive or negative, on how effective the response to tectonic mega-disasters is. Generally, response is less effective in developing countries like Haiti due to low incomes, weak governance and lack of preparation. However, mega-disasters can overwhelm even the most developed and prepared countries such as Japan during the 2011 earthquake and tsunami. Very high magnitude, geographically widespread mega-disasters can also cause major impacts — although in developed countries these tend to be in the form of economic losses rather than deaths. The extent to which a tectonic event can be predicted is also an important variable, meaning the impacts from earthquakes are usually more severe than the impacts from eruptions.

ⓔ **The student demonstrates accurate and relevant knowledge and understanding for AO1. They apply their knowledge and understanding to make relevant connections between development level, other relevant factors and impacts, supported by evidence and with sound judgements to create balanced and coherent arguments for AO2.** **Level 3, 11 marks**

Student B

(a)

State the eruption that caused the highest death toll	2006, Ecuador
Calculate the mean death toll	45
State the median value for death toll	5
Calculate the percentage of volcanic eruptions that occurred in Indonesia	4 out of 15

ⓔ **Two of the answers here are incorrect. The mean death toll is stated as 45, whereas the correct answer is 44.3. This error is likely to be a result of rushing, so errors creep in. The student also only gives a fraction (4 out of 15) for the calculation of the percentage of eruptions occurring in Indonesia, whereas a percentage is needed.** **2 marks**

(b) Mega-disasters occur whenever earthquakes or volcanoes cause impacts on people.

A lower level of development can affect the vulnerability of the local populations in a negative way. Unsustainable development can make local people more vulnerable. In Haiti, because of poverty, there was a lack of proper building design and poor construction techniques before the 2010 earthquake. As a result, many of the buildings were flattened as the floors collapsed on top of each other.

A lack of money can also result in under-investment, so that the population is not properly prepared to deal with hazard events. And a population is more vulnerable if there is a low level of social development and nutrition and health levels are lower to begin with. Following an earthquake, diseases can quickly spread through makeshift camps that people set up. In Haiti in 2010, for example, a cholera outbreak following the earthquake caused an additional 9,000 deaths. Lack of money also affected the rebuilding programme after the earthquake. By 2016, 60,000 of the 1.5 million Haitians left homeless after the quake were still living in camps.

In richer countries, development can have a positive effect. The Japanese have built extensive sea-wall defences against the tsunami threat. However, there is evidence that these walls led to complacency among the local population during the 2011 quake. That resulted in many of them ignoring the tsunami evacuation warnings until it was too late.

e **The student shows some accurate and generally relevant knowledge and understanding for AO1, although to improve their mark they should include more detail. The student has applied their knowledge and understanding to make some relevant connections between development and impacts, supported by evidence. Overall, however, more connections could be made and the answer is imbalanced between the positive and negative effects of development for AO2. The answer has limited assessment of the degree to which development level is important.** **Level 2, 5 marks**

Question 2 mark scheme

(a) 4 marks (AO3 = 4 marks)
  - State the prefecture which experienced the first tsunami waves = Miyagi prefecture
  - Calculate the range for the tsunami run-up heights = 32.2 m (37.9 m – 5.7 m)
  - Calculate the range for the tsunami run-up heights in Miyagi prefecture = 12.3 m (18 m – 5.7 m)
  - Calculate the mean tsunami run-up height for Iwate prefecture = 23.6 m

(b) 12 marks (AO1 = 3 marks, AO2 = 9 marks)
Some suggested ideas are given below but you may wish to expand on these or include other relevant points.

AO1 Demonstrating your knowledge and understanding

There are three main categories of response, each requiring increasing amounts of technology to achieve their goals.
  - Modifying the loss, e.g. through aid following a disaster.
  - Modifying the vulnerability, e.g. through prediction/warning, community preparedness.
  - Modifying the event, e.g. coastal defences against tsunamis, making buildings 'life safe'.

AO2 Applying your knowledge and understanding
  - Make relevant connections.
  - Support your evaluation with evidence.
  - Produce a balanced and coherent argument.

There are various ways in which you can modify loss (aid, insurance, etc.), vulnerability (prediction and warning, community preparedness and education) and the event itself (life-safe buildings, engineering defences such as tsunami walls). You could discuss these through the use of contrasting examples, making direct and clear connections between them to allow you to apply the three factors above to produce a balanced argument.

Answers to this question will be given a mark within a level band

Level 1 (1–4 marks) You show only a limited geographical knowledge and understanding of the factors affecting the success of different responses. You make limited connections between aspects of your answer and support your interpretations with limited evidence. You draw unbalanced conclusions based on the material in your answer.

Level 2 (5–8 marks) You show mostly relevant and accurate geographical knowledge and understanding of the factors affecting the success of different responses. You make mostly relevant connections between aspects of your answer as appropriate and support your interpretations with some evidence. You draw conclusions based on the material in your answer but your conclusions may be limited or unbalanced.

Level 3 (9–12 marks) You show relevant and accurate geographical knowledge and understanding of the factors affecting the success of different responses. You make sound connections between aspects of your answer as appropriate and support your interpretations logically with evidence. You draw balanced and logical conclusions based on the material in your answer.

Hints and tips

Consider the various factors that can affect the success of different responses, coming to an overall assessment at the end of your essay.

Question 2 example responses

Student A

(a)

State the prefecture which experienced the first tsunami waves	Miyagi
Calculate the range for the tsunami run-up heights	32.2
Calculate the range for the tsunami run-up heights in Miyagi prefecture	12.3
Calculate the mean tsunami run-up height for Iwate prefecture	23

ⓔ **The first three answers in the table are correct, but the last answer is not stated to one decimal place: the correct answer is 23.3, so this answer would not gain a mark.** **3 marks**

(b) Earthquakes, tsunamis and volcanic eruptions can be responded to in different ways. These include modification of the loss with strategies such as insurance cover against earthquake damage, and short- and long-term aid after a disaster has occurred. Second, modification of vulnerability through strategies such as prediction and warning, community preparedness, education on how to respond to the onset of a tectonic hazard and the aftermath of the event. Finally, modification of the event can be done through the use of technology to mitigate the impacts, including designing buildings to be life safe and the use of coastal engineering to defend against tsunamis.

ⓔ **The introduction provides a clear structure for the main body of the essay.**

The most effective response is modification of the event, however this is rarely possible. Earthquakes cannot be prevented, and even secondary hazards such as tsunamis are hard to stop. Japan's costly and technologically advanced tsunami walls were only partially successful in reducing damage from the 2011 Sendai tsunami: the death toll of 15,500 with US$360 billion in economic losses demonstrates the extent to which the tsunami defences were overwhelmed. In rare cases volcanic lava flows can be diverted, such as on Mount Etna in the 1980s, but in most cases they cannot be. Overall, event modification is usually impossible or impractically expensive — especially in developing countries. However, simple land-use zoning can be used to prevent people living in especially high-risk areas, e.g. areas of liquefaction risk or coastal zones at high tsunami risk. This is low cost, and if adhered to strictly can reduce the numbers in harm's way.

ⓔ **This section contains a good assessment, recognising that while event modification is desirable it is rarely applicable.**

Loss modification could be viewed as a failure in terms of response, not a success. In the aftermath of a tectonic disaster local and international aid often pours into an area for rescue, food and shelter, and to prevent the spread of disease. However, this response demonstrates the underlying vulnerability of people and their lack of local coping capacity and resilience. Billions in international aid were needed in the aftermath of the 2010 Haiti earthquake. In developed countries insurance covers losses in many cases, but claiming insurance and rebuilding are slow processes.

The most successful approach is modifying the vulnerability of communities, by building resilience. Volcanic eruptions, and to some extent tsunamis, can be predicted, allowing for warning and evacuation which saves lives (but not property). Community preparedness in terms of education, skills to help people build hazard-proof homes and training in how to respond during and after a disaster, can all help. Vulnerability modification can be low cost and be applied in developed and developing world contexts. Because it occurs before a disaster strikes it has the capacity to reduce impacts enormously.

ⓔ **A clear argument in favour of modifying vulnerability as the approach most likely to be successful.**

Overall, all three modification responses should be used, but they are not all applicable to all hazard types. Loss modification risks 'picking up' the pieces in a disaster situation, unless vulnerability modification has been used prior to the event — in which case the losses should have been minimised. Modifying the vulnerability is especially important in terms of earthquakes, which are the most common tectonic hazard and one that cannot be predicted, therefore reducing vulnerability by careful planning and preparation is likely to be the most successful approach.

ⓔ **The student shows accurate and relevant knowledge and understanding for AO1. They apply their knowledge and understanding to make relevant connections between different types of response, supported by evidence and with sound judgements to create balanced and coherent arguments for AO2.** **Level 3, 11 marks**

Student B

(a)

State the prefecture which experienced the first tsunami waves	Iwate prefecture
Calculate the range for the tsunami run-up heights	32.2
Calculate the range for the tsunami run-up heights in Miyagi prefecture	12.3
Calculate the mean tsunami run-up height for Iwate prefecture	21.3

ⓔ **The first answer given, Iwate prefecture, is incorrect because Miyagi is closest to the epicentre of the earthquake that generated the tsunami. Both answers calculating range are correct. In the last answer, the candidate's answer of 21.3 is incorrect because they have included data for Kesennuma in their calculation and this town is in Miyagi prefecture not Iwate.** **2 marks**

(b) When it comes to trying to successfully respond to earthquake hazards, there are a number of things that a country can do. For example, you can try to modify the event itself. You do this by using strategies to reduce the impacts, such as tsunami walls. You can modify people's vulnerability by training them in how to respond when an earthquake happens and by making buildings life safe. You can also respond to the earthquake when it happens, for example by providing tents for homeless people.

ⓔ **Some good knowledge and understanding shown of the relevant factors, but more detail could be given; answer is narrowly focused on earthquakes.**

Different countries at different levels of development can do these things to different extents. For example, the Haiti earthquake in 2010. Haiti is a very poor country at a low level of development. When a magnitude 7.0 earthquake struck, many of its buildings collapsed and up to 300,000 people were killed. This was because the country did not have proper earthquake plans in place beforehand, for example building codes. As a result, many buildings pancaked, killing the people inside. The last major earthquake in Haiti was in 1751, and so most people were largely unaware of the earthquake risk that they faced. Also, it did not have a plan to deal with the aftermath of the earthquake. In fact, it had to give over control of the aftermath management to the United Nations. In contrast, the following year in Japan a huge magnitude 9.0 earthquake struck, much larger than the one in Haiti. However, only 18,000 people were killed — still a huge number, but far less than Haiti. The Japanese authorities had tried to modify the event by introducing strict building codes — only 14 of the buildings damaged were due to seismic shaking. They had also trained their population on what to do when an earthquake strikes. They also had an earthquake early warning system in place which allowed them to send out warnings after the earthquake happened but before the shock waves arrived to allow people to duck for cover under a table (drop, cover, hold). And in the aftermath, they had a fast and extensive response to the damage caused by the tsunami.

ⓔ **The student shows accurate and generally relevant knowledge and understanding for AO1, but the answer relies too much on descriptive detail and lacks assessment, i.e. AO2 skills. The student applies their knowledge and understanding to make some relevant connections between responses and impacts, supported by some evidence. Overall, however, a clearer assessment of success is needed to gain AO2 marks.** **Level 2, 6 marks**

Topic 2 Landscape systems, processes and change

Option 2B Coastal landscapes and change

Question 3 mark scheme

(a) (i) 6 marks (AO1 = 3 marks, AO2 = 3 marks)

AO1 Demonstrating your knowledge and understanding
- Discordant coastlines are characterised by a series of headlands and bays.
- More resistant rock forms the headlands as it is less vulnerable to erosion from the sea by processes such as hydraulic action and abrasion.
- In contrast, bays occur where less resistant rock is found — erosion by the sea causes the coastline here to retreat more rapidly.

AO2 Applying your knowledge and understanding

➤ The varying geology of West Cork has contributed to the formation of the discordant coast here.
➤ Limestone is less resistant to erosion and so it has retreated more rapidly to form bays such as Bantry Bay.
➤ On the other hand, more resistant rock, including sandstone and mudstone, has resisted erosion and so forms the headlands.

Answers to this question will be given a mark within a level band

Level 1 (1–2 marks) You show limited geographical knowledge and understanding of the formation of the discordant coastline. You apply your knowledge and understanding of the formation of the discordant coastline with limited effect, making limited connections between aspects of your answer and supporting your interpretations with limited evidence.

Level 2 (3–4 marks) You show mostly relevant and accurate geographical knowledge and understanding of the formation of the discordant coastline. You apply your knowledge and understanding of the formation of the discordant coastline, making some connections between aspects of your answer as appropriate and supporting your interpretations with some evidence.

Level 3 (5–6 marks) You show relevant and accurate geographical knowledge and understanding of the formation of the discordant coastline throughout. You apply your knowledge and understanding of the formation of the discordant coastline throughout your answer, making sound connections between aspects of your answer as appropriate and supporting your interpretations logically with evidence.

Hints and tips

Give a clear explanation of the relevant processes here. Develop your points to show good depth of understanding.

(a) (ii) 6 marks (AO1 = 3 marks, AO2 = 3 marks)

AO1 Demonstrating your knowledge and understanding

➤ Where rock strata run parallel to the shoreline, we find concordant coasts.
➤ This means that, unlike discordant coasts, geology tends to vary less along a coastline.
➤ However, geological structure can create more complex patterns along these shorelines. For example, the sea can erode more quickly where a rock stratum has faults and/or joints, and can cause the rock type along the coastline to be breached, increasing erosion behind it.

AO2 Applying your knowledge and understanding

➤ In the case of the Lulworth Cove area, this appears to be exactly what has happened. The more resistant rock adjacent to the sea has been breached where faults or joints occurred.
➤ The rock running parallel behind it has clearly been less resistant to erosion and a series of bays has formed along the shore, such as Lulworth Cove.
➤ The extensive area of chalk behind these other rock types has resisted erosion, causing the bays to extend out laterally along the coastline.

Answers to this question will be given a mark within a level band

Level 1 (1–2 marks) You show limited geographical knowledge and understanding of how recessional rates have been affected by geology on this coastline. You apply your knowledge and understanding of how recessional rates have been affected by geology on this coastline with limited effect, making limited connections between aspects of your answer and supporting your interpretations with limited evidence.

Level 2 (3–4 marks) You show mostly relevant and accurate geographical knowledge and understanding of how recessional rates have been affected by geology on this coastline. You apply your knowledge and understanding of how recessional rates have been affected by geology on this coastline, making some connections between aspects of your answer as appropriate and supporting your interpretations with some evidence.

Level 3 (5–6 marks) You show relevant and accurate geographical knowledge and understanding of how recessional rates have been affected by geology on this coastline throughout. You apply your knowledge and understanding of how recessional rates have been affected by geology on this coastline throughout your answer, making sound connections between aspects of your answer as appropriate and supporting your interpretations logically with evidence.

Hints and tips

Give a clear explanation of the relevant processes here. Make sure you develop your points to show good depth of understanding.

(b) 8 marks (AO1 = 8 marks)

AO1 Demonstrating your knowledge and understanding

There are two main wave types: constructive and destructive.

Constructive waves have the following characteristics:
➤ Longer wave lengths and lower wave heights
➤ Spilling breakers with a strong swash and weak backwash
➤ They occur on more gently sloping beaches
➤ They are common during less stormy conditions, e.g. the summer

Destructive waves have the following characteristics:
➤ Shorter wave lengths and higher wave heights
➤ Plunging breakers with a weak swash and strong backwash
➤ They occur on more steeply sloping beaches
➤ They are common during stormy conditions, e.g. the winter

Beach morphology is affected by these waves:
➤ Constructive waves tend to move material from the offshore zone further up the beach. So, over time, they tend to increase beach gradient.
➤ Destructive waves tend to move material from the top of the beach and take it down towards the bottom of the beach. So, over time, they tend to decrease beach gradient.
➤ This means that there is an annual cycle in beach profile: steeper, lower beaches found in the winter, and gentler, higher beaches found in the summer.

Sediment profiles are also affected by these waves. Storm beaches are found at the top of the beaches, formed as coarse sediment is thrown up to the top of the beach by destructive waves during extreme storms. The middle section of the beach is usually sand. The offshore bar is formed as destructive waves carry sediment off the beach and deposit it offshore.

Answers to this question will be given a mark within a level band

Level 1 (1–2 marks) You show limited geographical knowledge and a narrow understanding of the effect of wave type on beach profiles. Part of your answer may be inaccurate or lack detail.

Level 2 (3–5 marks) You show mostly relevant geographical knowledge and understanding of the effect of wave type on beach profiles. Some parts of your answer are not fully developed.

Level 3 (6–8 marks) You show accurate and relevant geographical knowledge and understanding of the effect of wave type on beach profiles. Your answer is detailed and fully developed.

Hints and tips

'Explain' questions require you to show good understanding of how these wave types affect the shore — justify your points to demonstrate this understanding.

(c) 20 marks (AO1 = 5 marks, AO2 = 15 marks)
Some suggested ideas are given below but you may wish to expand on these or include other relevant points.

AO1 Demonstrating your knowledge and understanding
➤ Short-term coast flooding can be caused by storm surges linked to storms and cyclones, or even inundation by tsunamis.
➤ Tectonic events, such as major earthquakes, can submerge coastlines.
➤ Major deltas are sinking due to subsidence.
➤ Isostatic change, over thousands of years, is leading to both emergent and submergent coasts.
➤ Global climate change, caused by human activity, is causing sea level rise across the globe on a short timescale.

AO2 Applying your knowledge and understanding
➤ Long-term changes have significant impacts on coastal landscapes and landforms, whereas short-term flooding events have a smaller impact.
➤ However, significant risks to people and property result from short-term flooding by hazardous events.
➤ Global sea-level change as a result of global warming could be significant on timescales of several decades, putting millions at risk.
➤ Some coastlines have multiple processes occurring at the same time, leading to complex interactions and coastal situations which are difficult to manage.

Answers to this question will be given a mark within a level band

Level 1 (1–5 marks) You include isolated points of geographical knowledge and understanding of the impacts of short-term coastal flooding and long-term sea-level rise on coastal areas, with some errors and inaccuracies. You have not made connections from the question to points made. Your answer is incoherent and lacks relevant evidence to support ideas. Your argument is limited, with unbalanced points. Points that you make are concluded in a general manner, if at all.

Level 2 (6–10 marks) You make some points showing geographical knowledge and understanding of the impacts of short-term coastal flooding and long-term sea-level rise on coastal areas, some of which may be relevant. You make some inaccurate points. You apply some knowledge and understanding of the effect of long- and short-term processes of sea-level change on coasts but your ideas are not developed or may not be linked directly to the question. You use some evidence to support statements, which may answer only part of the question. You make a conclusion but this is drawn from often unbalanced ideas.

Level 3 (11–15 marks) You make generally relevant points showing geographical knowledge and understanding of the impacts of short-term coastal flooding and long-term sea-level rise on coastal areas. Your ideas are mostly accurate and some connections are made between ideas. You interpret the question well in general but there may be some gaps in the use of evidence to support points. You draw a conclusion that links to the arguments made but is not fully supported by evidence.

Level 4 (16–20 marks) You show good use of geographical knowledge and understanding of the impacts of short-term coastal flooding and long-term sea-level rise on coastal areas. You make a range of relevant points to create a coherent argument supported by appropriate evidence. You apply your knowledge well throughout. All points you make are linked to the question. You draw a good, well-balanced conclusion that links clearly to the evidence presented.

Hints and tips

Give balanced attention to both of the processes (short and long term) and reach an overall assessment of their relative impact.

Question 3 example responses

Student A

(a) (i) A discordant coastline is one where you find a series of headlands jutting out into the sea and bays eroded back into the land. These are found typically where you have different types of geology at a coastline and where these bands of different rock are aligned perpendicular with the coastline.

ⓔ **Good understanding shown of the nature of and conditions required for discordant coastlines.**

In West Cork, we can see that this is indeed the case. The main coastline runs from northwest to southeast and there is a series of different rock types (including purple mudstone, limestone and old red sandstone) running at right-angles to the shoreline.

ⓔ **The student has applied their knowledge to the specific location shown in the figure.**

Where you have different rock types, they can erode at different rates. The less resistant rock is eroded faster by coastal processes such as hydraulic action and abrasion. This causes them to retreat to form bays. Meanwhile, the more resistant rock does not retreat at the same rate due to erosion, and so it forms bays that jut out from the coastline.

ⓔ **The student shows good understanding of the processes at work to form discordant coasts.**

We can see this happening in West Cork. Evidently, the areas of limestone are more vulnerable to erosion. These areas in particular have retreated through erosion to form three main bays including Dunmanus Bay, Bantry Bay and Kenmare River bay. Bantry Bay has been eroded back over 30 km. In fact, it is likely that these large bays are so big because of additional flooding following the end of the last glacial period as sea levels rose and river valleys were flooded to form rias. In contrast, the sandstones and mudstones have not been eroded to anywhere near the same extent and they form a series of headlands sticking quite far out into the sea.

ⓔ **The student makes good use of the figure to apply general understanding to the context of West Cork.**

However, there is some evidence that the headlands themselves are not being subject to significant erosion. For example, Clear Island, a section of old red sandstone towards the southeast of the map, has been separated from the rest of the headland. This is because wave refraction around headlands tends to concentrate erosion there. Some form of weakness, possibly a fault, has allowed erosion to cut right through the headland here, forming the island.

(e) **Good application of understanding to this one particular aspect of the stimulus material.**

Level 3, 6 marks

(ii) Concordant coasts are often found where the geological strata run parallel to the coastline. In places like this, in contrast to discordant coasts, the geology tends not to vary as much as you move along the coastline. This might suggest that erosional rates would remain similar along the coast and that you would not find the headlands and bays you find at discordant coasts. However, this is not the case, due to other factors including geological structure. For instance, the rock strata running along the coast will have faults (major weaknesses in the rock) and joints (fractures in the rock). These are much more susceptible to erosion and can allow the coastline to retreat at certain points, forming headlands and bays.

(e) **Detailed and extensive understanding shown of the nature of this coastline and the role geological structure can play in its development.**

We can see these processes in action in the Lulworth Cove area. The Portland beds and Purbeck beds run parallel to the shoreline and are more resistant to erosion. However, there are various bays that have formed here, including Lulworth Cove and the more extensive Mupe Bay/Worbarrow Bay. These have most likely formed as a result of faults and joints in the Portland and Purbeck beds. These would have allowed erosion to be concentrated here and to break through these strata, beginning to erode the softer Wealden and Gault beds behind. It seems that the extensive stratum of chalk that runs behind all the other strata is more resistant to erosion as the bays do not extend back into this rock type. Rather, erosion seems to be occurring laterally along the Wealden and Gault beds, especially along the 1-km-wide bay at Mupe Bay/Worbarrow Bay.

(e) **The student applies their understanding effectively to the situation in the Lulworth Cove area, suggesting how the strata and geological structure have affected the development of this coastline.**

Level 3, 6 marks

(b) There are two main wave types that typically occur at different times of the year. The first is constructive waves. These are more common during the summer and are associated with less stormy conditions. These waves are characterised by longer wave lengths and so lower wave heights. When they reach the beach, their low wave heights mean that they have spilling breakers which have stronger swash and a weaker backwash.

In contrast, destructive waves are more common in the stormy conditions normally found in the winter. These have higher wave heights and so their wave lengths are shorter. As they break on the beach, the taller waves tend to produce plunging breakers which break more vertically. This gives them less forward momentum and so they have a weaker swash and a stronger backwash.

This annual variation in wave type causes beach morphology to vary annually too. The plunging breakers and strong backwash of the destructive waves remove sediment from the top of the beach and carry it offshore. This not only lowers the level of the beach overall, it also makes the profile less steep. As the spring and summer come around, however, constructive waves with their spilling breakers and stronger swash tend to move sediment back onto and up the beach. This increases beach height and slowly increases its gradient too. So, we have an annual cycle of changing beach profile linked to different waves dominating at different times of the year.

These waves also influence the beach sediment profiles. The most severe destructive waves during the winter throw the coarsest sediment up to the top of the beach, forming a storm beach made up of shingle. Only the destructive waves have enough energy to move the largest beach material. The finer material including sand is found further down the beach, including the offshore bar which is a deposit of sand carried off the beach by the strong backwash of the destructive waves.

ⓔ **This is a detailed and comprehensive answer that shows accurate and relevant knowledge and understanding throughout.** **Level 3, 8 marks**

(c) Short-term coastal flooding can be caused by hurricanes, if low-lying coastal areas are vulnerable to the impacts of a storm surge. This occurs due to the extreme low pressures found in hurricanes and other storms. As a result of the reduction in the weight of the atmosphere above it, sea level may rise significantly. Hurricane Katrina in 2005 raised the sea level along the Louisiana coastline by close to 10 m. This then can cause severe coastal flooding, which happened in New Orleans. Flooding here resulted as the levées designed to protect the city failed and the storm surge was able to enter the city. This was further compounded by the topography of New Orleans. Much of the city is below sea level, and once this water entered the city it stayed there, only being pumped out nearly a week after the hurricane struck. Louisiana's Gulf Coast is also sinking as a result of subsidence of the Mississippi Delta, which makes the impact of short-term flooding worse over time. In addition, global sea-level rise due to global warming is increasing flood risk over time. Like many locations, New Orleans is experiencing complex interactions because short-term flooding — and its management — is being made worse by longer-term subsidence and sea-level rise as a result of global warming.

ⓔ **The student demonstrates a good grasp of the connections between various factors that affected storm surge flooding in New Orleans, and the complex interactions resulting from different long- and short-term changes.**

Short-term change in sea level may also result from tectonic processes. For example, the magnitude 9.0 earthquake that struck off the coast of Japan in 2011 lowered the level of the land in the Sendai Plain. Before the earthquake, around 740 acres were below sea level on the Plain — this rose to nearly 4,000 acres after the quake. This resulted from the land suddenly rebounding because of the sudden movement of the plate at the destructive margin to the east of the country. As the Eurasian Plate and the Pacific Plate collided, compression forces beyond the margin caused the Japanese coastline to rise slightly (like a crumple zone in the front of a car being compressed if it collides with a brick wall). However, when these tectonic pressures were released, the compression force was also released, allowing the coastline to sink back down again slightly. The impact of this change will be that these low-lying areas are much more at risk of flooding from the likes of typhoon storm surges, but also the longer-term impact of the expected 0.7–1 m sea-level rise expected by 2100 as a result of global warming.

ⓔ **The student shows a good understanding of how tectonic forces can produce changes in sea level and has explained well their impact on Japan.**

Sea level may change on longer timescales. One of the factors that produces these changes is the change from glacial to interglacial periods producing isostatic (local) and eustatic (global) changes in sea level. When the world enters a period of more intense glaciation, eustatic sea-level fall can occur as more and more of the Earth's water is locked up in continental ice. When we then enter an interglacial (as we did nearly 12,000 years ago), the melting of the glacial ice can add more water to the seas and oceans, producing a eustatic rise in sea level. However, this rise can be affected by isostatic changes in local regions. For example, immediately following the last glacial period in the British Isles, sea levels rose eustatically. At the same time, the melting of the ice on the land removed a significant weight and caused the land to rebound upwards. So, as the sea levels rose, the land rose too. The relative rates of these movements is key as to whether or not a stretch of coastline experiences rising or falling sea levels. At first, the eustatic rise was faster than the isostatic sea level fall. But, today, many parts of the British Isles are still rebounding isostatically. The area around Glasgow is rising at the rate of around 1.5 mm per year. At the same time, the southeast of England is sinking very slowly. It was not glaciated during the last glaciation, and as a result the whole northern part of the British Isles is springing upwards, while the southeastern portion is pivoting downwards slightly. These long-term changes have a significant impact on coastal landscapes, producing emergent coastlines with raised beaches and fossil cliffs, and submergent coastlines with rias and fjords. However, long-term changes occur on timescales of thousands of years so their impact on humans is minimal. On the other hand, the gradual subsidence of London and the southeast of England as result of isostatic readjustment does add to the problem of sea-level rise as a result of global warming.

ⓔ **The student shows a well-developed grasp of the complex interactions that result from glacial and interglacial sea-level change and has supported interpretations with examples of evidence. They recognise that change on different timescales has different impacts, i.e. the physical landscape versus the impact on humans.**

In conclusion, short-term coastal flooding is caused by storm surges, and less commonly by tectonic processes. These changes have significant impacts on people because they occur in the context of natural disasters. These can be managed, but at significant cost, e.g. levées, flood defences, tsunami walls. Longer-term change as a result of glacial and interglacial cycles has a significant landscape impact, but a smaller impact on people. However, contemporary sea-level rise as a result of global warming is interacting with both short-term hazards and, to a lesser extent, longer-term processes such as subsidence. This means many coasts experience a complex set of interactions as sea level changes due to multiple causes.

ⓔ **A coherent conclusion following on from strong evaluation in the main section of the essay, drawing together the various strands of the assessment. The answer is well supported by evidence, and makes clear judgements recognising the complexity of the situation in terms of human and landscape impacts.** **Level 4, 20 marks**

Student B

(a) (i) A discordant coast is found where you have different types of geology at a coastline. Harder rock is found next to softer rock. The rock types meet the coast at an angle and so geology can be variable depending on where you are at the coast.

ⓔ **Some understanding of the characteristics of the discordant coast is shown, but the student should discuss the series of headlands and bays that results.**

You can see these different rock types in West Cork. The limestone strata have clearly been eroded back much further inland to form bays at places like Dunmanus Bay. But the sections of more resistant rock have not been eroded as much and so stick out and form headlands. Other bays include Bantry Bay and Kenmare River. These have been eroded by processes such as hydraulic action and abrasion to create these large bays.

ⓔ **The student attempts to apply their understanding to the figure, but the answer requires much more detail in terms of understanding of the processes at work here. In addition, the connections between the stimulus material and the question need to be drawn out more. The answer could be more clearly organised and structured.** **Level 2, 3 marks**

(ii) Where rock strata are parallel to the coastline, you have concordant coasts. You can still find headlands and bays here, however, because of other factors that can help them develop. These include geological structure. If there are areas of weakness in the rock, known as faults, these can erode more quickly and the sea can carve out bays in the softer rock behind.

ⓔ **The student demonstrates some good and relevant knowledge here, but it could be developed in more detail to show fuller understanding.**

In the Lulworth Cove region, we find bands of rock running parallel to the shoreline. These consist of rocks such as Portland beds and Purbeck beds. These are more resistant rocks, but they have faults in them. These faults have eroded over time to form bays here such as Lulworth Cove. Once the sea has eroded through the Portland and Purbeck beds, it erodes out the softer Wealden and Gault beds behind. This process carved out the bay found at Lulworth Cove.

ⓔ **Again, there is some good application here to the figure, but the answer only makes some of the connections between the stimulus material and the question needed to reach Level 3.** **Level 2, 4 marks**

(b) There are two main wave types common at different times of the year. The first is constructive waves, which are more common during the summer and are associated with less stormy conditions. These waves have longer wave lengths and so lower wave heights. Second, destructive waves are more typical in the stormy conditions of winter. These have higher wave heights and shorter wave lengths.

ⓔ **The student outlines some of the key features of the two waves, but does not include other important elements, such as references to swash and backwash.**

As a result, beach morphology can change over the course of the year. During the winter, beaches tend to be lower as the destructive waves move sediment off the beach. This also tends to make them become less steep. However, during the following summer, the beaches get their sediment back again as the constructive waves sweep it back onto the beach. The beach sediment profiles are also affected by these waves. Destructive waves move the largest sediment to the top of the beach during storms, creating a storm beach of shingle at the top. The rest of the beach is made up mostly of sand, carried onto the beach from the offshore bar by the constructive waves.

ⓔ **The answer is relevant to the question asked, but the ideas are not explored in enough detail. As a result, the answer is underdeveloped.** **Level 2, 4 marks**

(c) Short-term changes are the result of tides. In addition to the usual daily tides, sea level can vary a bit more over the course of a month — these are called spring or neap tides. Spring tides occur when the moon and sun are aligned with each other and produce slightly greater tidal ranges. Neap tides occur 7 days after spring tides and occur when the moon is at right angles to the sun. The gravitational pull is slightly less, so tidal ranges are slightly lower. One of the coastal landforms that can be affected by tides is mudflats. These form when the tide moves in and out more slowly and where you have a shallow offshore gradient. Spring tides can also lead to problems of coastal flooding. For example, in January 2017, a spring tide corresponded with strong winds in the North Sea to cause flooding in eastern England. The high tides and strong winds funnelled the water south towards mainland Europe. As a result, 5,000 homes on the east coast of England were evacuated and lowland in north Norfolk flooded.

ⓔ **This is a detailed descriptive section on tides, with minimal evaluation.**

Tropical storms can also cause storm surges because of the extreme low pressure found in these storms. For example, the sea level rose by up to 10 metres in the storm surge produced by Hurricane Katrina in 2005. This inundated the city of New Orleans as the levées burst and the city was flooded. As the city lies below sea level, this water was unable to drain away. Hurricane Katrina was one of the worst natural disasters to affect the USA in modern times.

Tectonic processes can also affect sea levels in the short term. When earthquakes occur at destructive margins, the land can suddenly sink or rise up. In Japan in the 2011 earthquake, sections of the coast sank down a bit. This allowed the tsunami to overtop some of the flood defences, and has left the country more vulnerable to tsunamis in the future. In contrast, an earthquake in New Zealand in 2016 caused the sea bed to rise by 2 metres, lifting up some previously submerged landforms above sea level along parts of the coast.

ⓔ **This section has some good AO1 knowledge and understanding, but less evaluation.**

Sea levels can also change over longer time periods. As our planet has moved between glacials and interglacials during the Pleistocene and into the Holocene, sea levels have risen and fallen. Global rises in sea level are called eustatic rises, and local rises are called isostatic rises. Eustatic rises occur when ice that was on the land melts and enters the sea. Isostatic falls also occur when the ice melts — but in this case the weight of the ice coming off the land causes it to rebound upwards, lifting the land 'out of' the sea and causing sea levels in effect to fall. These two changes interact in a complex way. Sea level can be rising eustatically while at the same time falling isostatically. You can end up with complex coastal landforms as a result. If the sea level is falling, then former coastal features can now be found raised up above the sea and a few hundred metres inland, for example raised beaches and fossil cliffs. If it is rising, then the sea can flood inland, forming features like rias (flooded river valleys) and fjords (flooded U-shaped valleys in glacial areas).

ⓔ **There is a range of causes of sea-level change and flooding, but in an answer that describes and explains but largely fails to evaluate. AO1 is quite strong, but AO2 is much weaker. There is limited use of evaluative language and no conclusion, so the answer is limited to Level 2.**

Level 2, 10 marks

Question 4 mark scheme

(a) (i)   6 marks (AO1 = 3 marks, AO2 = 3 marks)
Some suggested ideas are given below but you may wish to expand on these or include other relevant points.

AO1 Demonstrating your knowledge and understanding
➤ Waves are refracted around the headland, concentrating erosion here.
➤ Erosion can take place by three main processes: hydraulic action, corrasion/abrasion and corrosion.

AO2 Applying your knowledge and understanding
➤ These erosional processes will preferentially erode places of weakness in the rock, such as a joint, bedding plane or fault.
➤ Where the fault runs perpendicular to the coastline, caves can be eroded, which may ultimately become blow holes and geos.
➤ Where the faults run parallel to the shoreline, through the headland, the caves that form there can ultimately become arches, stacks and stumps.

Answers to this question will be given a mark within a level band

Level 1 (1–2 marks) You show limited geographical knowledge and understanding of the sequence of coastal processes and resultant landscape. You apply your knowledge and understanding of the sequence of coastal processes and resultant landforms with limited effect, making limited connections between aspects of your answer and supporting your interpretations with limited evidence.

Level 2 (3–4 marks) You show mostly relevant and accurate geographical knowledge and understanding of the sequence of coastal processes and resultant landscape. You apply your knowledge and understanding of the sequence of coastal processes and resultant landforms, making some connections between aspects of your answer as appropriate and supporting your interpretations with some evidence.

Level 3 (5–6 marks) You show relevant and accurate geographical knowledge and understanding of the sequence of coastal processes and resultant landscape throughout. You apply your knowledge and understanding of the sequence of coastal processes and resultant landforms throughout your answer, making sound connections between aspects of your answer as appropriate and supporting your interpretations logically with evidence.

Hints and tips

Give a clear explanation of the relevant processes here. Develop your points to show good depth of understanding.

(a) (ii) 6 marks (AO1 = 3 marks, AO2 = 3 marks)
Some suggested ideas are given below but you may wish to expand on these or include other relevant points.

AO1 Demonstrating your knowledge and understanding

➤ Longshore drift is an important transportational process at coasts.
➤ It occurs when waves approach a shoreline at an angle. The swash moves up the beach at an angle, carrying sediment with it. The backwash comes back down straight, moving the sediment further along the shore with it. The sediment is moved along the beach in a zig-zag pattern.
➤ Deposition takes place in areas where wave energy drops, leading to a build-up of sediment which can be re-worked over time by new processes.
➤ The landforms created are spits, cuspate forelands and recurved spits.

AO2 Applying your knowledge and understanding

➤ The way longshore drift interacts with a coastline is determined by a combination of prevailing wind(s) and coastal orientation.
➤ Where a coast changes orientation suddenly (e.g. at an estuary), longshore drift carries on in the original direction, forming a spit (if the spit grows over the whole estuary mouth it forms a bar).
➤ Where two longshore drift currents approach each other from opposite directions, the waves cancel each other out, forming a triangular feature of deposition called a cuspate foreland.
➤ Secondary prevailing winds and/or offshore currents can work alongside the dominant prevailing wind to form a recurved spit.

Answers to this question will be given a mark within a level band

Level 1 (1–2 marks) You show limited geographical knowledge and understanding of the role longshore drift plays in the formation of the landforms. You apply your knowledge and understanding of the role longshore drift plays in the formation of the landforms with limited effect, making limited connections between aspects of your answer and supporting your interpretations with limited evidence.

Level 2 (3–4 marks) You show mostly relevant and accurate geographical knowledge and understanding of the role longshore drift plays in the formation of the landforms. You apply your knowledge and understanding of the role longshore drift plays in the formation of the landforms, making some connections between aspects of your answer as appropriate and supporting your interpretations with some evidence.

Level 3 (5–6 marks) You show relevant and accurate geographical knowledge and understanding of the role longshore drift plays in the formation of the landforms throughout. You apply your knowledge and understanding of the role longshore drift plays in the formation of the landforms throughout your answer, making sound connections between aspects of your answer as appropriate and supporting your interpretations logically with evidence.

Hints and tips

Give a clear explanation of the role of longshore drift as it relates to the various landforms. Make sure you develop your points to show good depth of understanding.

(b) 8 marks (AO1 = 8 marks)

Some suggested ideas are given below but you may wish to expand on these or include other relevant points.

AO1 Demonstrating your knowledge and understanding

➤ Sand dunes need various conditions for their formation, including a supply of sand, onshore winds, a large area of flat land behind the shore.

➤ However, it is the stabilising role that vegetation plays that is key in their development.

➤ As sand starts to build up around debris towards the backshore, it begins to be colonised by plants such as sea couch. These plants bind the sand together, stabilising its height and allowing it to build up further from the influence of the sea.

➤ This encourages more and different plants to grow — these not only continue to bind the sand together with their roots, but they also add organic matter to the soil, improving its structure and making it less susceptible to wind erosion.

➤ As soil quality improves further, percentage vegetation cover increases and the soil is further protected from erosion from the wind.

➤ However, the stabilising actions of the vegetation in sand dunes can be adversely affected by various human activities, such as trampling and sand extraction.

Answers to this question will be given a mark within a level band

Level 1 (1–2 marks) You show limited geographical knowledge and a narrow understanding of the role that vegetation plays in dune successional development. Part of your answer may be inaccurate or lack detail.

Level 2 (3–5 marks) You show mostly relevant geographical knowledge and understanding of the role that vegetation plays in dune successional development. Some parts of your answer are not fully developed.

Level 3 (6–8 marks) You show accurate and relevant geographical knowledge and understanding of the role that vegetation plays in dune successional development. Your answer is detailed and fully developed.

Hints and tips

'Explain' questions require you to show good understanding of the specific role of vegetation — justify your points to demonstrate this understanding.

(c) 20 marks (AO1 = 5 marks, AO2 = 15 marks)

Some suggested ideas are given below but you may wish to expand on these or include other relevant points.

AO1 Demonstrating your knowledge and understanding

➤ Integrated Coastal Zone Management (ICZM) is often used to manage coastlines in a holistic way.
➤ Shoreline Management Plans (SMPs) are used as the framework for decision-making for coastal management in the UK, as part of ICZM.
➤ There are four broad policy approaches to choose from: hold the line; advance the line; managed retreat; no active intervention.
➤ Cost–benefit analysis and Environmental Impact Assessment (EIA) are important aspects of decision-making.
➤ Both hard and soft engineering approaches can be used to manage coastlines.

AO2 Applying your knowledge and understanding

➤ Decisions involve evaluating a range of factors, including the value of land and property to be protected, but 'value' could be in non-monetary terms and is difficult to establish.
➤ There may be environmental reasons behind decisions which outweigh economic considerations.
➤ SMPs require impacts on the physical environment and the human environment (land use, tourism/recreation, heritage, communities) to be considered, leading to complex decisions which may be controversial and not satisfy all players.
➤ Any specific policy decision can be controversial: there can be losses and losers, and these may be sacrifices for greater overall benefits.

Answers to this question will be given a mark within a level band

Level 1 (1–5 marks) You include isolated points of geographical knowledge and understanding of the decision-making factors, with some errors and inaccuracies. You have not made connections from the question to points made. Your answer is incoherent and lacks relevant evidence to support ideas. Your argument is limited, with unbalanced points. Points that you make are concluded in a general manner, if at all.

Level 2 (6–10 marks) You make some points showing geographical knowledge and understanding of the decision-making factors, some of which may be relevant. You make some inaccurate points. You apply some of your knowledge, but your ideas are not developed or may not be linked directly to the question. You use some evidence to support statements, which may answer only part of the question. You make a conclusion but this is drawn from often unbalanced ideas.

Level 3 (11–15 marks) You make generally relevant points showing geographical knowledge and understanding of the decision-making factors. Your ideas are mostly accurate and some connections are made between ideas. You interpret the question well in general but there may be some gaps in the use of evidence to support points. You draw a conclusion that links to the arguments made but is not fully supported by evidence.

Level 4 (16–20 marks) You show good use of geographical knowledge and understanding of the decision-making factors. You make a range of relevant points to create a coherent argument supported by appropriate evidence. You apply your knowledge well throughout. All points you make are linked to the question. You draw a good, well-balanced conclusion that links clearly to the evidence presented.

Hints and tips

Give balanced attention to various considerations that go into an SMP and reach an overall assessment of their relative impact.

Question 4 example responses

Student A

(a) (i) At headlands, waves are refracted as they approach the headland, bending around it and concentrating their energy here. This means that headlands are places that experience significant erosion. This can occur by three main methods: hydraulic action (waves break on the shore and trap air in crevices in the rock, increasing air pressure, putting stress on the rock); abrasion/corrasion (sediment carried by the waves being thrown against the shore by the breaking waves); and corrosion (carbonic acids in the water dissolving limestone and chalk).

🄮 **The student shows a strong understanding of these processes and has communicated them clearly.**

The way these erosional processes operate at a coastline is influenced by various factors, especially the geology. The landforms shown in the diagram are typical in places with sedimentary rocks such as sandstone and limestone with bedding planes and joints.

🄮 **The student outlines clearly the factors that are relevant to the formation of the features in the figure.**

In places where rocks are faulted, erosion is concentrated along these planes of weakness. Where the faults run parallel to the main shoreline, running through the headland, a sequence of processes occurs. First, a notch is carved out by erosion, widening out into a cave which can cut through the entire headland, forming an arch. The unsupported roof of this cave collapses to form a stack (the stack eventually is eroded away leaving only a stump). Where the joins run perpendicular to the main shoreline, the cave erodes back into the land. First, the water may be funnelled back into the cave to erode a blow hole in the ceiling. Eventually, the entire cave roof collapses to form a narrow inlet called a geo.

🄮 **The student applies their understanding of the erosional processes and factors to the particular scenario outlined in the figure.** **Level 3, 6 marks**

(ii) Longshore drift is an important process of transportation along coastal areas. As the wave breaks at an angle, the swash moves up the beach at the same angle. Any sand or shingle carried by the breaking wave thus moves up the beach at the same angle. However, the backwash follows the most efficient way back down the beach, and so moves back down straight. Again, this carries the sediment back with it. The end result is that the sediment moves along the beach with each breaking wave, following a zig-zag route.

ⓔ **A clear understanding of the operation of longshore drift.**

How this process actually interacts with a coastline is determined mostly by the prevailing wind (the most commonly occurring wind direction) and the orientation of the coast. For example, when a coastline changes orientation suddenly (such as at an estuary mouth), the prevailing wind carries the longshore drift out across the estuary mouth. This forms a feature known as a spit. The end of the spit can form a hook as a secondary prevailing wind causes currents to curve the end round.

When you have a place where two prevailing winds cause longshore drift to operate in opposite directions, where the currents meet they can cancel each other out and cause deposition to occur. This can form a triangular-shaped feature called a cuspate foreland.

ⓔ **The student clearly applies their understanding to the various depositional features shown in the figure, explaining the role longshore drift plays in their formation.** **Level 3, 6 marks**

(b) Sand dunes may form where there is a large supply of sand and sediment (e.g. from a beach), where there are onshore winds for periods of time and where there is a large area of flatter land behind the shore. However, the role of vegetation in stabilising the sand during sand dune plant succession is vital in the formation of a stable dune system.

Sand found above the high-tide level is transported by onshore winds towards the back of the beach. Where there are obstacles such as pieces of driftwood or seaweed, the sand can build up in height, forming the embryo dunes. This increased height raises the beach up slightly away from the influence of the tides and allows plant succession to begin. Salt-tolerant plants such as sea couch colonise the embryo dunes. Their roots help stabilise the sand, and further deposition occurs around the long, thin leaves of the grass, further raising the height of the dunes and forming the foredunes. Foredunes can build quickly as plants like marram grass begin to grow, forming a height of several metres in 5–10 years. As they grow, they improve the soil quality further as they add organic matter to the soil. This improves its structure and further stabilises the soil, making it much less susceptible to wind erosion. Furthermore, the better soil quality allows a wider variety of plants to grow, including rosette plants (e.g. dandelions) and the percentage vegetation cover in the dunes rises to 100%. This protects the soil from further wind erosion and increases its stability.

The importance of vegetation in stabilising sand dunes can be seen when the vegetation is removed by erosion. This can occur naturally, forming large hollowed-out bowls of sand called blowouts. But it can also happen because of human activities. Sandy beaches and dunes are popular recreational destinations. As people trample through the dunes, the vegetation can die and wind erosion can degrade the dunes considerably. Many dune systems in the UK are now managed to try to mitigate the impacts of trampling on dune stability.

ⓔ **The answer shows a good understanding of the various ways in which vegetation in sand dunes can act to stabilise the dunes. The reference to human trampling to show what happens when vegetation is removed is an effective illustration of the importance of vegetation.**
Level 3, 8 marks

(c) Land value, and wider economic considerations, are important aspects of decision-making on coastlines that suffer from coastal recession of high flood risk, but there are many other factors in what is normally a complex decision. These include environmental issues, technical feasibility, i.e. whether defences are actually possible, as well as social issues. In the UK, the overarching aim is that coastlines are managed in a holistic, sustainable way using the principles of Integrated Coastal Zone Management (ICZM) with the framework of Shoreline Management Plans (SMPs). SMPs manage large stretches of coastline, based on physical sediment cells, in order to avoid piecemeal management that can do more harm than good.

In some locations land value is a key consideration. Hornsea on the Holderness Coast is protected from erosion and coastal flooding by a combination of flood walls, sea-walls, rip-rap and a groyne field. These classic hard-engineering solutions implement a 'hold the line' policy that protects the town, fishing port and even Hornsea Mere, a national nature reserve. The cost–benefit analysis of spending £15 million on Hornsea's defences is very positive. However, land value can be reduced in importance. In the early 1990s about £2 million was spent protecting the hamlet of Mappleton a few kilometres downdrift of Hornsea, despite the settlement being valued at only £650,000 at the time. Political pressure, applied by angry locals to the local council and government, ensured a 'hold the line' policy was implemented. Along most of the Holderness Coast the policy is 'do nothing' because land at risk is low-value farmland. Mappleton is, in that sense, an anomaly.

In some cases, environmental factors are the most important consideration. The Blackwater Estuary in Essex was one of the first locations in the UK to be managed by 'strategic retreat'. Existing flood barriers were breached, allowing seasonal and storm flooding to take place. This was done partly because of the low value of the land (grazing land) but also because it was felt existing flood defences could not be sustained because of the long-term risk from rising sea levels. There was an additional environmental benefit of creating new areas of valuable marsh habitat, so arguably the key decisions made were environmental ones.

As SMPs work to manage the whole of a coastline, sometimes difficult decisions have to made about protecting some areas or not. This is the case at Happisburgh near Cromer in north Norfolk. The SMP recommends managed retreat as the policy for this section of coastline. The SMP assessment of the impact on the physical environment of this policy states that erosion will allow continued exposure of the 6 hectares of cliffs with Site of Special Scientific Interest status at Happisburgh. Elsewhere in the sediment cell, the sediment eroded from these cliffs will contribute to beaches and dunes further to the south, maintaining and enhancing the dunes at places like Eccles-on-Sea and Winterton-on-Sea. Furthermore, there will be loss of Grade 1 agricultural land to erosion from the sea, totalling up to 45 hectares by 2105.

In its assessment of the impacts on the human environment, the SMP states there will be a significant loss of commercial and residential properties as a result of the managed retreat option. By 2025, around 15 properties in Happisburgh will have been lost, primarily along Beach Road. In addition, there will be a loss of land from the clifftop caravan park and the HM Coastguard rescue facility. By 2050, it is expected that cumulative losses of properties will be between 15 and 20. There will be further loss of land from the clifftop caravan park at Happisburgh. By this stage, the Grade I St Mary's church and the Grade II Manor House will be at risk of erosion. By 2105, total residential losses are expected to be between 20 and 35 properties and the SMP foresees the probable loss of St Mary's church and the Manor House.

Despite these impacts, the policy of managed retreat is sustainable into the longer term. The loss of land will be a major input of sediment into the sediment cell, and it will be transported south along the coast to enhance the beaches and dunes found there. This will not only protect settlements such as Eccles, but it will also help to maintain and develop natural habitats. It will also allow the coastline to operate in a natural manner. So, despite the loss of residential properties, cultural heritage and economic assets, these losses are not sufficient to justify the costs of protecting the coastline here, never mind the negative impact such interventions would have in other areas to the south.

In conclusion, we can clearly see how complex the range of interacting factors are that must be considered when making policy decisions for coastal management. While land value is clearly important, it is by no means the only factor in decision-making. The integrity of wider management aims and long-term sustainability are the main factors at Happisburgh, whereas at Mappleton political and social considerations proved more important than economic ones. No one factor is dominant.

ⓔ This is a detailed and comprehensive answer. The student shows wide-ranging knowledge of SMPs and the decision-making process, and understanding of how different factors are linked together and inform each other in the decision-making process. The use of examples from north Norfolk and elsewhere gives the answer useful evidence to support the argument.

Level 4, 19 marks

Student B

(a) (i) The features shown here are erosional. There are various erosional processes at coasts: hydraulic action (waves break on the shore and trap air in crevices in the rock, increasing air pressure, putting stress on the rock); abrasion/corrasion (sediment carried by the waves being thrown against the shore by the breaking waves); and corrosion (carbonic acids in the water dissolving limestone and chalk).

ⓔ The student demonstrates a good grasp of the erosional processes.

At a headland, a notch is carved out by erosion. This can widen out into a cave and this can cut through the entire headland to form an arch. As the arch is unsupported by material below, it will eventually collapse to leave an isolated stump offshore. Elsewhere, the caves can erode inland by wave action. As the water travels down the cave, it can break upwards towards the end and erode a hole in the roof (the blow hole). In fact, the entire roof may collapse leaving an inlet called a geo.

ⓔ The student shows a good understanding of how the processes create the landforms. To gain more marks, the student could set this in the context of the factors that affect how these processes operate, especially geology. **Level 2, 4 marks**

(ii) As the wave breaks at an angle, the swash moves up the beach at an angle, carrying sand or pebbles with it. However, the backwash moves back down straight, again this carries the sediment back with it. So, the sediment moves along the beach with each breaking wave, in a zig-zag route. This process is known as longshore drift.

ⓔ The student demonstrates a solid understanding of the operation of longshore drift.

At an estuary mouth, longshore drift occurs out across the estuary mouth. This forms a feature known as a spit. If the spit builds across the entire estuary mouth, it forms a bar. When two longshore drift currents move in opposite directions, where they meet they can cancel each other out and cause deposition to occur. This can form a triangle-shaped feature called a cuspate foreland.

ⓔ The student outlines the processes that operate to form these features. The answer would be improved by exploring the factors that affect how longshore drift operates (such as prevailing wind and coastal orientation) and by explaining the specific role of longshore drift in the formation of a tombolo. **Level 2, 3 marks**

(b) Sand at the back of the beach starts to build up around obstacles such as pieces of driftwood. As it builds up, plants can begin to colonise, e.g. sea couch. These plants have rhizome roots that spread out laterally and help to bind the sand together. And the leaves of the plants encourage more sand to build up, improving stability.

As the dunes get higher, other plants start to grow too, for example marram grass. This adds more organic matter to the soil and this means that it binds together better and is more stable. When the winds blow over the dunes, they are less likely to erode the sand because of this, so the dunes are more stable.

We can see just how important the vegetation is in dune stability when we think about how people can destroy the vegetation by walking over it. When this happens, the dunes can be badly eroded. Many sand dune systems in the UK need to be managed to help fight against this.

(e) **The student shows a good overall grasp of most of the relevant material here, but the answer needs to show greater depth of understanding to gain top-level marks.** **Level 2, 4 marks**

(c) Land value is the most important factor when it comes to coastal management. This is because the value of the land influences what can be justified in terms of strategy and defences against coastal flooding and erosion. Where land values are high, the usual policy is 'hold the line'. This is the case with high-value infrastructure such as the Easington Gas Terminal on the Holderness Coast, which is protected by groynes and rip-rap. Close by, most of the land is just farmland, caravan sites and farms. This is low value, and so it is not protected and the policy used is 'do nothing'. A cost–benefit analysis can be done on new coastal defence plans to show the economic case for protecting land, or not protecting it. This is why millions was spent in the 1980s on things like the Thames Barrier to protect London from coastal flooding. The value of London is so high that the decision to protect it was easy.

(e) **The answer uses examples and has some details, but the argument is one-sided — despite showing some understanding of coastal management and policy choices.**

At Newbiggin-by-the-Sea in Northumberland about £10 million was spent on new sea defences. This included a sea wall, an offshore breakwater and beach nourishment. These were designed to prevent future erosion and protect Newbiggin and the shops, houses and other businesses there. The value of Newbiggin was important in this decision and other areas that are vulnerable but not valuable were not protected. Also at Newbiggin there was an attempt to improve the beach and promenade to boost tourism, so the sea defences had a wider purpose than just protecting from erosion and flooding. It was a multi-use scheme. Specifically the beach nourishment was decided on to make the beach look more attractive for visitors.

(e) **There is some, brief, evaluation here as the answer starts to hint that other factors beyond the land value are important.**

Overall, land value is the main factor because how much land is worth determines how much money can be used to protect it from erosion and flooding. This can be seen at Newbiggin, Easington and even London.

(e) **This answer is one-sided, it gives one view rather than evaluating a number of different views. There is an attempt at a conclusion, but it is really just a summary of what has been said. The evaluation in the answer is very brief.** **Level 3, 11 marks**

Area 2 Dynamic places

Topic 3 Globalisation

Question 1 mark scheme

(a) 4 marks (AO1 = 4 marks)

This question focuses on demonstrating your knowledge and understanding of part of the globalisation process (technological change). You gain 1 mark for giving a reason why globalisation has been accelerated by ICT. The remaining 3 marks can be earned by developing the initial reason using examples and additional linked explanations. For example, how different types of technology have accelerated different parts of the globalisation process, e.g. social, cultural change.

For example:

ICT and mobile communications enable citizens, businesses and states to interact with one another quickly (1) and over a 24-hour and 7-day-a-week time period (1). This allows information and ideas to be spread faster over large distances and across borders (1). ICT and mobile technology benefit TNCs (1) as businesses can grow their markets into new countries more easily (1) and improve the efficiency of their operations (1). The use of the internet has led to the spread of different cultural (1) and political ideas (1) to new audiences (1) and allows family and friends to keep in contact over long distances (1). Mobile phone technology has also allowed cheaper ways (1) of connecting people and services where, in the past, development has been held back because of the high cost of infrastructure (1).

Other appropriate reasons will be accepted.

Hints and tips

Think about the ways in which the internet is used by people and businesses and how this might lead to faster globalisation of services and ideas.

(b) 12 marks (AO1 = 3 marks, AO2 = 9 marks)

This question asks you to apply your knowledge and understanding of how the growth of economies can have impacts on a place's population and environment. You must use examples such as China and India to assess the social and environmental changes that come from rapid economic growth. As the question asks 'to what extent', a good answer must consider examples of positive aspects of economic change. Relevant points that you could mention and expand on with examples are suggested below.

AO1 Demonstrating your knowledge and understanding

➤ There has been a global shift of economic activity from Europe and North America to emerging economies, such as those in Asia.
➤ The development of manufacturing and services has resulted in changes to the built environment.
➤ This has led to benefits and costs for emerging economy societies and changes and challenges for the environment.

AO2 Applying your knowledge and understanding

➤ Unplanned settlement and planned settlement growth can result in different social and environmental challenges, e.g. water, sanitation, loss of community.

> Increased employment and wages have led to a rise in standards of living and disposable income for some workers but low wages and low employment security for some groups.

> Poverty has been reduced in some regions and for some groups of people, while the cost of living has risen for others.

> Looser environmental controls can lead to increased pollution and the loss of green space.

Answers to this question will be given a mark within a level band

Level 1 (1–4 marks) You show only a limited knowledge and understanding of the social and/or environmental challenges of economic growth in emerging economies. The examples you use to support statements may be very general and lack case study detail. At the bottom of this band you show no attempt to assess the challenges or put forward possible benefits.

Level 2 (5–8 marks) You generally show a good understanding of the social and environmental challenges of economic growth in emerging economies. The examples you use to support statements are relevant. You make some attempt to assess the challenges and to put forward some social and environmental benefits of rapid economic growth. You start to consider how impacts may vary between different places and different groups.

Level 3 (9–12 marks) You show accurate knowledge and understanding of the social and environmental challenges of economic growth in emerging economies. You apply your knowledge to effectively assess a range of issues and to make a judgement as to the effect these have on a range of places and groups of people. Benefits as well as challenges are examined. At the upper end of this band, you confidently use detailed case study information in your supporting statements.

Hints and tips

Which emerging economies can you use in your answer? What are the costs and benefits of rapid economic growth? How is society affected? How is the environment affected?

Question 1 example responses

Student A

(a) ICT and mobile communication have contributed to the acceleration of globalisation because <u>information can be exchanged more quickly and over greater distances</u> than before. The increase in percentage internet usage in many countries in the world has enabled more <u>families that live apart to keep in touch more easily</u>, e.g. by using Skype. The increase in internet bandwidth has enabled TNCs, such as HSBC, to <u>conduct their global business more quickly and to take advantage</u> of the 24-hour banking economy. Higher mobile phone accessibility and capability have also enabled migrants to send money home more securely. Mobile communication networks in developing countries, e.g. <u>Ethio telecom in Ethiopia,</u> have been developed at a cheaper cost than traditional telephone networks, allowing more connectivity and access to global markets and ideas.

(e) **A good answer giving clear reasons for the acceleration of globalisation, linked clearly to ICT and mobile communication. The student mentions different types of globalisation (social, cultural) and uses named examples well to justify their points.** **4 marks**

'information can be exchanged more quickly and over greater distances' Clear reason given.

'families that live apart to keep in touch more easily' Clear link between technology and social aspect of globalisation.

'conduct their global business more quickly and to take advantage' Link between use of ICT in global business.

'Ethio telecom in Ethiopia' Example to support answer.

(b) The rapid growth of emerging economies, particularly since 2000, has seen significant changes for both societies and the places in which they live. The development of both the manufacturing and service economies has led to challenges for workers but has also brought benefits and opportunities. The fast-paced development of urban environments has also led to a range of environmental issues which will need to be addressed if continued growth is to be sustainable.

Social challenges include the rapid growth of unplanned settlements, e.g. Mumbai's Dharavi slum in which low-income families live in overcrowded conditions with poor access to clean water, adequate sanitation and a regular electricity supply. Planned housing developments can lead to the demolition of traditional neighbourhoods and the movement of communities, e.g. the Hutongs, Beijing. However, rises in per capita income, e.g. in China from US$300 to US$7,500 between 1990 and 2015, show an increase in employee wages, allowing them to have more disposable income for goods such as cars and smartphones. However, in some economic sectors, e.g. textiles, wages can be low and the working day long, leading to employee exploitation. Permits may also be required to work, e.g. shoe manufacturing in Jiangsu, restricting the movement of workers. As GDP growth increases, governments may also choose to spend more money on education and training to improve the flexibility of their future workforce.

Environmental challenges include the loss of green space through urban development. For example, Chongming Island north of Shanghai has been developed to accommodate the need for space for 660,000 residents. The lack of parks and recreational spaces in new high-rise developments also reduces the environmental quality of cities. Air pollution has risen where increased car ownership and traffic congestion, e.g. in Beijing, has contributed to the rise in sulphur dioxide and particulate matter. However, the development of public transport systems, e.g. Shanghai's metro system (14 lines built in 20 years), have led to improvements in traffic flow, helping to reduce pollution.

While economic change in Asia has brought opportunities, it is clear that challenges to both society and the environment remain. These issues will need to be addressed if future growth is to be sustainable.

ⓔ **The student shows a good range of knowledge of different social (unplanned settlements, loss of communities, low wages) and environmental (loss of green space, air pollution) challenges, and applies knowledge well to appropriate case study examples. The answer is balanced by discussion of positive effects of economic growth (higher disposable income, investment in education, development of public transport). Specific named examples support the main points and a judgement is made to suggest that some challenges still need to be overcome. Examples are mainly from China; supporting evidence from other emerging economies would show a broader application of knowledge.** **Level 3, 11 marks**

Student B

(a) Mobile phones allow people to <u>communicate with their friends and family all over the world</u> and at any time of the day. They can transfer money from their bank account and catch up with social media. ICT allows goods and services to be <u>transferred quickly from one country to another</u>.

ⓔ The student understands the question but gives general factual statements that do not clearly link the use of ICT and mobile communication with the acceleration of globalisation and therefore do not receive credit. The student makes social and economic points, but clearly identified examples would improve the answer, for example 'social media, e.g. Facebook, YouTube'. The final point on 'goods and services' could be linked to TNCs and expanding markets.   **2 marks**

'communicate with their friends and family all over the world' Clear link with social aspect of globalisation.

'transferred quickly from one country to another' Speed of transactions linked to globalisation.

(b) Fast growth in Asia has led to many problems. In Beijing, pollution from cars is very high. This means that traffic congestion is also a problem. Lots of green areas have been built on and wildlife has lost its habitats. Also in Beijing, many homes have been destroyed to build new ones for the large number of people who are coming to live and work there. The number of factories in China has increased. More people can come from rural areas to work there, often for low wages. In Mumbai, many people live in slums as they cannot afford good houses. Many people live in very poor conditions and do not have access to clean water or proper sanitation. They often have low-skilled jobs which do not pay high wages. They can be exploited easily.

There have been some benefits to economic growth. People have a higher income and so more people are able to buy a car. Some people can afford to live in better quality houses. Some people may be able to find a better job in an office, like a call centre or a bank. The government may provide better services for its people, for example schools and hospitals. The reputation of the country may improve and it can host world events such as the Olympics in Beijing.

ⓔ The student identifies some social and environmental challenges. The supporting statements are sometimes linked to examples but these are often generalised. There is some attempt at assessment as benefits of economic growth are mentioned, but again this section lacks a range of specific examples needed to achieve Level 3. Marks could be gained by making a judgement as to whether the challenges are greater than the benefits (or vice versa).

Level 2, 7 marks

Question 2 mark scheme

(a) 4 marks (AO1 = 4 marks)

This question asks you to show your knowledge and understanding of the benefits and costs of local sourcing (obtaining products, usually food stuffs, from the nearby area). You gain 1 mark for identifying a valid reason(s) why local sourcing by local groups and NGOs brings benefits to a community, up to a maximum of 2 marks for a clear explanation of one reason or identification of two reasons. You gain 1 mark for identifying a valid reason(s) why local sourcing by local groups and NGOs brings costs to a community, up to a maximum of 2 marks for a clear explanation of one reason or identification of two reasons.

For example:

Locally sourced goods are those that are both produced and sold in the nearby area. While it is seen to be a sustainable approach to production, it can bring both benefits and costs. Farmers can attract higher prices for their food by producing it organically (1). They can also focus on specialities to sell in the local area (1). Produce grown locally and sold at nearby markets can reduce transport costs (1), which can help the farmer to increase their profit, particularly at times when fuel prices are high (1). Their carbon footprint may also be reduced (1). However, there are also disadvantages linked to local sourcing. These types of products are often more expensive than food bought from supermarkets (1). This is because the supermarket has a larger buying power and can offer its products at a lower price to the consumer. Also, increased sales of local products may mean that people are less likely to buy goods sourced from developing countries, e.g. cut flowers from Kenya (1). This may affect the development of these countries.

Other appropriate responses will be accepted.

(b) 12 marks (AO1 = 3 marks, AO2 = 9 marks)

AO1 Demonstrating your knowledge and understanding

➤ Global hub cities are those that are highly connected to the rest of the world.
➤ Global hub cities attract a range of international migrants with different levels of education, skill and influence.
➤ International migration has increased in global hub cities.

AO2 Applying your knowledge and understanding

➤ There are many impacts, both benefits and costs, of having international migrants in a global hub city.
➤ These impacts can be for the migrants themselves and for the host city.
➤ These impacts can be economic (e.g. linked to employment, income, taxation), social (e.g. linked to access to standard of living, housing and services, education), political (influence, policy change) and environmental (e.g. space, quality of surroundings).
➤ Some impacts are likely to be more significant than others, such as large-scale immigration placing pressure on housing and services, or the positive economic impacts of immigration in the UAE.

Answers to this question will be given a mark within a level band

Level 1 (1–4 marks) You show only a limited knowledge and understanding of the impacts of international migration within global hub cities. The examples you use to support your statements may be very general and lack case study detail. At the bottom of this band you show no attempt to assess the benefits and costs of international migration within these types of cities.

Level 2 (5–8 marks) You generally show a good understanding of the impacts of international migration within global hub cities. The examples you use to support statements are relevant. These impacts may be economic, social, political or environmental. You make some attempt to assess the benefits and to put forward some costs of international migration within these cities. You start to consider how impacts may vary between different cities and/or different groups.

Level 3 (9–12 marks) You show accurate knowledge and understanding of the impact of international migration within global hub cities. You apply your knowledge to effectively assess a range of benefits and costs (economic, social, political and environmental) and to make a judgement of the effect these impacts have on a range of cities and groups of people. At the upper end of this band, you confidently use detailed case study information in your supporting statements.

Hints and tips

Think about the impacts, both benefits and costs, that international migration has within global hub cities. Can you think of economic, social, political and environmental impacts? Are these positive or negative? Refer to specific global hub cities you have studied to add detail to your answer. You need to make a judgement about which impacts have been the most significant.

Question 2 example responses

Student A

(a) The production and selling of products which are grown or made locally can bring many advantages to the local community itself. Fruit, vegetables and meat are often produced by local farmers and the local community can support their business by buying these products. Some farmers enhance their crops by lowering the amount of pesticides that they use, which has the benefit of making healthier products. Farmers can also market their goods as organic if they follow the right procedures and this can add value to the price that they can charge. The community can also work towards being self-sufficient and so can move towards a better level of food security. Fewer journeys to markets further away will lower fuel bills. This also helps reduce their carbon footprint. However, local products can be more expensive and so may not be as affordable for people. Farmers may also find it difficult for their products to compete with larger supermarkets that can offer similar products for a lower price.

ℯ **This is a good answer. The student has given a wide range of economic, social and environmental benefits and has justified the points clearly. Two specific costs of locally produced products are given. There is good structure to the answer and good use of relevant vocabulary. 4 marks**

'support their business by buying these products' Benefit: Profit and supplying local markets.

'making healthier products' Benefit: Better diets and quality of products.

'add value to the price that they can charge' Benefit: Profits for local farmers.

'better level of food security' Benefit: Food security.

'lower fuel bills' Benefit: Farmers' overall income.

'reduce their carbon footprint' Benefit: Environmental sustainability.

'so may not be as affordable' Cost: High price of product.

'difficult for their products to compete' Cost: Economic viability.

(b) Global hub cities are those that are highly connected to the rest of the world through business and trade. They will also have social and cultural links to places in other countries. International migration, the permanent or semi-permanent movement of people from one country to another, has increased in these types of cities as people seek opportunities that they may not be able to find so easily in their home country. On the one hand, highly educated migrants may come to cities such as London. As London is a global financial centre, those migrants with banking skills, as well as entrepreneurs who are interested in start-ups, may come to develop their careers and to earn a higher income than they would in their own country. On the other hand, some global hubs attract lower-waged international migrants to fill the employment gaps left by the local population. For example, in Dubai many thousands of Indian workers work in the construction industry. They are able to send remittances home to India in order to support their families there. Also, the presence of international migrants in these cities can lead to increased tax revenue for the host country, income for local landlords and also create a cosmopolitan atmosphere in different districts, e.g. Brick Lane in London. However, international migrants can have a negative impact on global hub cities. For example, in London house prices in certain areas can rise as landlords can charge more rent for higher earners, some of whom have their rent paid for by their TNC employer as they have had to move abroad. Also, for lower-income international migrants, the quality of their accommodation may be poor and their tenancy may be linked to their job. If they lose their job, they may lose their home. Also, some international migrants work illegally, e.g. in cleaning and catering within London, and so have limited rights and do not make tax or national insurance contributions.

The rise in international migrants has brought both benefits and costs for both migrants and the global hub cities themselves. For richer sections of the population, the advantages may be greater than those for lower-income groups.

ⓔ **A good answer that addresses both the benefits and costs, focusing on economic (wages, type of work, taxes) and social (housing) impacts; referring to at least two social groups (high- and low-income workers); and giving two examples (London and Dubai). For higher marks, a broader range of impacts could be given, e.g. political influence of international migrants on decision-making (EU employees in Brussels); environmental impacts (increase in carbon footprint of migrants and families/friends travelling for visits). Values would add depth to the points, e.g. approximate numbers of migrants, amount of remittances, etc.**

Level 3, 10 marks

Student B

(a) Farmers like to supply their local area with goods as they can sell them to local people easily as they do not have to spend much money on petrol to transport the goods. They may also produce things which are not available in other places and so people may come to them to buy these products. One type of product which is popular is organic vegetables. These vegetables are grown in soil which does not have pesticides. There are strict rules about how you can grow these products and how you can market them. If farmers are successful with how they grow their crops, they can charge a higher price and make more profit for themselves.

ⓔ **This is a promising answer, but the student focuses solely on the benefits of locally sourced products and does not give any costs. Although at least three good benefits to the local community are given, a maximum of 2 marks can be awarded for these. The question specifically asks for costs also, but as the student does not address this no marks can be given.**

2 marks

'do not have to spend much money on petrol' Benefit: Cost of the product.

'not available in other places' Benefit: Growth of speciality products.

'make more profit' Benefit: Profit for the farmer.

(b) Global hub cities are the home of many people who have moved abroad to work. Some migrants from poorer countries like India move to the UAE to find jobs in the construction industry. This is because local people do not want to work in these types of jobs and so there is a shortage of workers. People come here and then send money home to their families. This is good for both the city, as it helps to get things built, and for the worker, as they earn more than they would in their own country. They can send money back home to help support their family. However, sometimes these types of workers are not treated well and may live in poor conditions. Some workers in cities like London may have come over illegally and this means that they do not have good employment rights and may be vulnerable. However, some international migrants come to global hub cities because of the buzz that is created there. Some young graduates may find it easier to develop their business there for the global market because the law makes it easier to set up and run their business. Some migrants may also be attracted by the fact that English is a common business language in some global cities and may find it easier to settle and find work as they have English as their second language.

ⓔ **The student shows some understanding of the impacts. Benefits/costs focus on economic aspects. The discussion about the 'buzz' of a global city needs to be related directly to the question (cosmopolitan atmosphere, wider range of products/services). More social (e.g. effect on school place and housing availability), political (e.g. change in host-government policy as response to increased international migration) and environmental impacts (e.g. carbon footprint increases) could be included. The student attempts to assess the benefits and costs, but no final judgement is made.** **Level 2, 7 marks**

Topic 4 Shaping places

Option 4A Regenerating places

Question 3 mark scheme

(a) (i) 3 marks (AO1 = 2 marks, AO2 = 1 mark)

You gain 1 mark for analysing the resource and recognising that there has been an overall increase in the percentage of homeworkers in rural villages. A further 2 marks can be gained by expanding the reasons for this change, up to a maximum of 3 marks.

For example:

The percentage of homeworkers in rural villages has increased to 25% during this period, which may reflect the improvement in internet access in rural villages (1). Increased internet availability and faster connectivity make it easier for people in some rural areas to work more flexibly (1) and to choose where they spend their working day (1).

Other appropriate responses will be accepted.

Hints and tips

You must make sure you select the correct line on the graph. Can you identify the change? Why has the change taken place? Don't forget to use percentages from the graph to help explain the overall trend.

....

(ii) 6 marks (AO1 = 3 marks, AO2 = 3 marks)

This question examines the reasons why there is a difference in the number of homeworkers in urban and rural areas. Relevant content that could be included is suggested below. Not all of these points would be needed for maximum marks. Alternative relevant suggestions will be credited.

AO1 Demonstrating your knowledge and understanding

➤ There is a clear difference between the percentage of homeworkers within urban and rural areas.
➤ More homeworkers live in rural areas: villages (25%) and hamlets (33%).
➤ Only 12% of homeworkers live in urban areas or towns in rural areas (16%).

AO2 Applying your knowledge and understanding

➤ Improvements in internet connections which make homeworking more viable.
➤ An increase in people working in the knowledge-based economy who can do business over the internet.
➤ These people may wish to have a better work–life balance and to work in a calmer environment than the city.

Answers to this question will be given a mark within a level band

Level 1 (1–2 marks) You show some general geographical knowledge and understanding of why there are differences between the percentage of homeworkers in urban and rural areas but some of this is inaccurate. You apply your knowledge to geographical information inconsistently.

Level 2 (3–4 marks) You show mostly relevant geographical knowledge of why there are differences between the percentage of homeworkers in urban and rural areas. In general, you apply your knowledge and understanding relevantly and find connections between the source material and the question.

Level 3 (5–6 marks) You show accurate and relevant geographical knowledge and understanding of why there are differences between the percentage of homeworkers in urban and rural areas throughout. You apply this knowledge and understanding logically to find fully relevant connections between the source material and the question.

Hints and tips

How might differences such as technology, transport, lifestyle choice and type of employment affect the location of homeworkers?

....

(b) 6 marks (AO1 = 6 marks)

This question focuses on why rural areas may have changed their demographic characteristics (such as age structure and ethnic composition) over time. Suggested ideas are outlined below, but you do not need to include all of these in your response. Other relevant points will be given credit.

AO1 Demonstrating your knowledge and understanding

➤ Some rural areas have suffered from depopulation as younger people move away to study and/or find employment.

➤ Some rural areas may have increased their population as new groups of people find village life desirable or find specific types of employment there.

➤ Some rural settlements may experience an ageing population.

Answers to this question will be given a mark within a level band

Level 1 (1–2 marks) You show isolated geographical knowledge and a narrow understanding of why rural places may have changed their demographic characteristics over time. Part of your answer may be inaccurate or lack detail.

Level 2 (3–4 marks) You show mostly relevant geographical knowledge and understanding of why rural places may have changed their demographic characteristics over time. Some of your geographical ideas are not fully developed.

Level 3 (5–6 marks) You show accurate and relevant geographical knowledge and understanding of why rural places may have changed their demographic characteristics over time. Your ideas are detailed and developed fully.

Hints and tips

Focus on rural areas. Think about how demographic characteristics such as age, gender and ethnicity would be relevant. Which examples will you choose to develop your answer?

(c) 20 marks (AO1 = 5 marks, AO2 = 15 marks)

This question focuses on how the regeneration strategies used in urban areas may be assessed in different ways by different groups. A range of stakeholders, such as local and national governments, local businesses, residents and community groups, are involved in the regeneration process and may have different criteria for success. The views that they hold may be formed by the experience that they have of living and working in the urban place. They may also be formed by the way strategies have an impact on the built environments and communities in reality and/or through perception.

Suggested ideas are outlined below, but you do not need to include all of these in your response. Other relevant points will be given credit.

AO1 Demonstrating your knowledge and understanding

➤ Urban regeneration aims to improve the built and/or social environment for those living and working within it.

➤ Strategies used to regenerate urban areas come from a variety of different stakeholders (interested parties).

➤ Different criteria are used by different groups to judge the success of urban regeneration strategies.

AO2 Applying your knowledge and understanding

➤ The success of regeneration will depend on the engagement of different stakeholders.

➤ The perception of success may be influenced by different factors, such as personal experience, age, gender, ethnicity, motives and media coverage.

➤ Urban regeneration strategies involving national governments may be linked to longer-term planning goals or political policies, whereas local councils assess the needs of their area and prioritise projects depending on social, economic and environmental requirements.

➤ Private developers may be concerned with the profit gained from urban regeneration, whereas local residents may have different viewpoints depending on how they will be affected by the regeneration.

Answers to this question will be given a mark within a level band

Level 1 (1–5 marks) You show isolated points of geographical knowledge and understanding of how the success of regeneration strategies can be viewed differently by different stakeholders, with some errors and inaccuracies. You show limited understanding and are not able to make connections. Your answer is incoherent and lacks relevant evidence to support your ideas. There is limited argument, and your points are unbalanced. Your ideas are concluded in a general manner, if at all.

Level 2 (6–10 marks) You make some points showing knowledge and understanding of how the success of regeneration strategies can be viewed differently by different stakeholders, some of which may be relevant. You make some inaccurate points. You apply some of your knowledge, but your ideas are not developed or may not be linked to the question. You use some evidence to support statements that may answer only part of the question. You give a conclusion but this is drawn from often unbalanced ideas.

Level 3 (11–15 marks) You show geographical knowledge and understanding of how the success of regeneration strategies can be viewed differently by different stakeholders. Your ideas are mostly relevant to the question and you make accurate points. You make some connections between your ideas. You interpret the question well in general, but there may be some gaps in how you use evidence to support points. You draw a conclusion that links to the arguments made but may not be fully supported by evidence.

Level 4 (16–20 marks) You show good use of geographical knowledge and understanding of how the success of regeneration strategies can be viewed differently by different stakeholders. You make a range of relevant points to create a coherent argument supported by relevant evidence. All of your points link to the question. You draw a good, well-balanced conclusion that links clearly to the evidence presented.

Hints and tips

Who is involved in regeneration? Why might different groups view regeneration differently? Use specific, named urban regeneration strategies to help explain your points.

Question 3 example responses

Student A

(a) (i) There has been <u>a rise in the percentage of homeworkers</u> in rural villages from 21% in 2006 to 25% in 2016. This increase reflects the fact that more people are able to work from home because of <u>better internet connections</u> in rural areas. More rural areas have broadband networks and improved wifi, allowing people <u>to work more flexibly in a more peaceful, greener environment</u>.

ℯ **The student has written a good answer that clearly identifies the correct trend and puts forward a valid reason for the increase in homeworkers in rural villages.** **3 marks**

ℯ **'a rise in the percentage of homeworkers' Accurately identified trend for rural villages.**

ℯ **'better internet connections' Valid reason for increase given.**

ℯ **'to work more flexibly in a more peaceful, greener environment' Expansion of given reason.**

(ii) Figure 1 shows a clear difference between the percentage of homeworkers in more populated settlements (e.g. urban areas and rural towns) <u>than in smaller, more remote settlements</u>. The number of homeworkers in the more populated settlements is just above or below the average for England (<u>14%</u>), whereas homeworkers number <u>25% in rural villages and 33% in rural hamlets</u>. This is because some homeworkers choose to move away from the city to enjoy <u>a quieter lifestyle</u>. If they have a job which they can do using the internet, better infrastructure links such as fibre optic cables <u>have improved broadband services to remoter areas</u>. This allows them to work from home. However, some workers may have <u>jobs which cannot only be done using the internet</u>, e.g. those working in large factories, in hospitals or schools. These types of employment are often found in settlements with large populations.

ℯ **The student makes good use of the source material to identify differences, including using evidence from the graph (percentages). They apply geographical knowledge well to the question using relevant information to support their points.** **Level 3, 6 marks**

'than in smaller, more remote settlements' Key difference identified.

'(14%)' Difference supported by percentages from the graph.

'25% in rural villages and 33% in rural hamlets' Difference supported by percentages from the graph.

'a quieter lifestyle' Valid reason.

'have improved broadband services to remoter areas' Reason clearly linked to technological changes.

'jobs which cannot only be done using the internet' Reason clearly linked to employment types.

(b) The <u>movement of people both in and out of rural areas</u> has led to changes in the population structure of rural settlements. Remote villages, e.g. in central Wales, are at risk of <u>losing young workers</u> who move away to go to university or to find employment. Jobs here in the agricultural and tourism sectors are more limited and not so well paid. East Devon <u>attracts more wealthy retired people</u> as they may move to be closer to families, as they are sometimes involved in childcare for grandchildren. <u>Younger families then find it difficult to stay</u> in the area as house prices increase. This leads to a higher proportion of elderly people in these areas.

Some smaller towns in more rural regions, which have traditionally been composed of mainly white British residents, have had an <u>increase in migrants from Eastern Europe,</u> reflected in a diversity of services such as the Polish shop in Huntingdon. Some rural areas such as villages within the London commuter belt, e.g. Penshurst in Kent, may have had a slight increase in population over time as they may <u>attract wealthier residents with jobs in the quaternary sector,</u> e.g. finance.

ⓔ **The student gives a good answer with several supported reasons for demographic change in rural areas. The characteristics put forward (age, employment type, ethnicity) are appropriate and are linked to relevant examples.** **Level 3, 6 marks**

'movement of people both in and out of rural areas' Reason given for change in population.

'losing young workers' Reason for change of age structure given with example.

'attracts more wealthy retired people' Reason for change of age structure given with example.

'Younger families then find it difficult to stay' Reason supported with relevant statement.

'increase in migrants from Eastern Europe' Reason for change linked to ethnicity with example.

'attract wealthier residents with jobs in the quaternary sector' Reason supported by change of employment type with examples.

(c) The regeneration of urban areas involves many groups of people. Stakeholders have different interests and motivations when regeneration takes place and may be affected in different ways. These varying viewpoints have an effect on how the groups perceive the success of a project.

The London Olympic Park was built as part of a range of venues hosting the 2012 Olympic Games. It is an example of sport-led regeneration. The brownfield site along the Lea Valley in East London was developed by a range of different players including the Olympic Delivery Authority (a government-appointed body with responsibility for the park's infrastructure), and local London boroughs (e.g. Newham and the Greater London Authority, responsible for transport improvements). Other groups involved included local businesses located in the proposed park, such as Forman's salmon smokehouse, and residents from the Clay's Lane housing co-operative, a residential development set up by Newham Borough Council. The investment in the project was £9.3 billion and it was hoped the development would have a positive impact on the built environment and the economy of the local area and for local communities. However, different organisations viewed the level of success differently.

In general, the regeneration was seen as a success by the government, raising the profile of the UK and injecting £13 billion into the national economy. The GLA helped to develop new Underground links, increasing the area's connectivity with west London, and the new Westfield Stratford City shopping centre satisfied investors with its high retail revenues, although this has benefited national rather than local retail chains. Residents of the new eco-housing development, East Village in the Olympic Village, were pleased to move to new homes, and a school (Chobham Academy) was built, providing improved education opportunities for local children.

However, some local groups saw the change more negatively. Forman's factory had to be moved and the company was concerned that this would affect its business. Some local residents were angry because Clay's Lane was demolished, particularly as this had provided subsidised accommodation in an area with high housing costs. Local projects that had been involved in training young people in construction skills to help them find employment before and after the Games were unhappy, as many of those trained found it difficult to get work and so the scheme was perceived by some as unsuccessful. West Ham United relocated to the main Olympic Arena, now the London Stadium. The football club sees this positively as it can increase the size of its revenue and boost the club's profile. However, some fans were not happy about the move away from their traditional home. Some environmental groups were concerned about the loss of green space, but the Lea Valley Regional Park has improved access to the river area for local residents.

The Olympic regeneration example shows that the level of success of a project is seen differently by people in the public, private and voluntary sectors. While a large-scale project, such as the Olympic Park site, is often the catalyst for economic growth in an area, there are some groups who feel that they will be disadvantaged in the long term, both economically and socially. Success can be judged more easily in economic terms (e.g. through revenue and profit) than in other ways (e.g. through satisfaction and opportunity).

e **The student uses a range of strategies from one case study. Economic, social and environmental strategies are discussed and both positive and negative viewpoints highlighted. The conclusion uses some of the arguments from the main essay. Good geographical knowledge is shown and evidence relating to the question is applied well. A balanced conclusion is made. The student could analyse the question further by using a contrasting example of urban regeneration, e.g. a small-scale, local project. They could also have approached the essay by focusing on the viewpoints of each group in turn, illustrating differences using examples of urban strategies they have studied.** **Level 4, 17 marks**

Student B

(a) (i) The number of homeworkers has increased because they have <u>better access to the internet</u>.

ⓔ **The student does not clearly link the increase to rural villages in their response. A valid reason is given but this could be expanded further, e.g. mentioning why rural villages have had better internet connections.** **1 mark**

'better access to the internet' Valid reason given.

(ii) In 2013, the number of people who were working at home was higher in rural areas than urban areas. The <u>highest number was those working in rural hamlets and dispersed</u> settlements. This is because people who live in the countryside are able to work at home more because they can <u>use the internet</u>. In urban areas fewer people work from home as the <u>internet is less common</u>. They may not need to use the internet for their job. Some people in urban areas may have <u>jobs in a place that they have to go to each day</u>, such as a shop or a leisure centre.

ⓔ **In general, the student understands the question although there are some inaccuracies. They try to apply knowledge but there is limited expansion of their points.** **Level 2, 3 marks**

'highest number was those working in rural hamlets and dispersed' The key point of the graph is understood; percentages could have been included to support the answer.

'use the internet' True, but the same can be said for urban areas, so this point would need expanding for more marks, e.g. improved connectivity, wifi.

'internet is less common' This is an inaccurate reason — the student has misinterpreted the meaning of the graph.

'jobs in a place that they have to go to each day' Links reason to employment type.

(b) The population of rural areas has changed in many ways over time. <u>Fewer people are employed</u> in farming and now work in tourist jobs. People have moved out of villages as it can be difficult to find a job there. In more popular villages, house prices may be high and so <u>people there who want bigger houses will have to move away</u>. Villages also <u>may be seen as boring</u> by people, with nothing to do there and bad transport links with other places. <u>Rich people</u> can buy houses instead and can work from home some of the time. Some <u>elderly people</u> like to retire to villages as they feel safer and more secure.

ⓔ **The student makes some good points but does not provide examples of places to support their answer. Their reasons are limited (mainly linked to employment) and they should discuss more fully the changes in the age structure of rural places.** **Level 2, 3 marks**

'Fewer people are employed' The points linked to employment type are general with no expansion.

'people there who want bigger houses will have to move away' Could be linked to age, e.g. people with young families, to make a clearer link back to the question.

'may be seen as boring' This statement is not linked to the question.

'Rich people' Link to a change in the wealth and occupation of residents.

'elderly people' Starts to put forward age as a characteristic that has changed but more expansion is needed.

(c) Different groups of people can have different opinions on urban projects. Parts of a city can be improved by building better housing for people. Housing is important as people need somewhere to live and often houses in cities are expensive and are not good quality. BedZED is an environmentally friendly housing project in south London. It aims to reduce the emission of greenhouse gases and water. Residents can also share cars. Homes are heated by solar panels and are well insulated. The scheme is supported by a charity, and is built to be eco-friendly. It also promotes One Planet Living, which tries to encourage people to get most of what they want locally and to promote sustainability. The houses are either sold, rented or under 'shared ownership'. The scheme also works with local businesses to help them to make their produce more sustainable. Trafalgar Place in south London is a housing development built by a company. It is part of the redevelopment of the Elephant and Castle area. The buildings are energy efficient and local residents have green spaces to enjoy. There are wheelchair-accessible homes too.

People can view the success of regeneration projects differently. Businesses and investors want to make a profit and may be looking for short-term success. Local residents may have a mixed view. If they think that they will be badly affected by the project, i.e. losing their home or dealing with increased traffic, they will not support the scheme. If they feel that they will benefit from the regeneration, they may feel more positively. Volunteer organisations may raise awareness about parts of a project that other groups may ignore. For example, environmental groups may complain about the loss of green areas and children's charities may protest against the loss of play space.

The success of an urban regeneration scheme depends on the attitudes of the people involved. Also, what some groups may perceive as success, others may not. Groups that are involved with the project judge the benefits and costs of a project differently.

ⓔ **The student shows some good knowledge of case study examples but these do not always link to the question. Different groups are mentioned, but there is no explanation of how they judge the success of the projects differently. Contrasting examples examining different types and scales of regeneration, in different locations, would help the student to discuss the role of a more varied range of interest groups and to evaluate how they view success. The second part of the response makes some connections to the question, but the points are not clearly linked to the groups mentioned in the first section. A conclusion is given, although it is brief and general, and there is limited evaluation.** **Level 3, 12 marks**

Question 4 mark scheme

(a) (i) 3 marks (AO1 = 2 marks, AO2 = 1 mark)

You gain 1 mark for suggesting one reason for the difference between the number of men and women working part-time. A further 2 marks can be earned for expanding the reason for this, up to a maximum of 3 marks.

For example:

There are more women (6.18 million) working part-time than men (2.25 million) in the UK (1). This may be because some women require a more flexible approach to working (1) in order to accommodate responsibilities for either children or elderly relatives (1). Many women are also employed in the caring and leisure sectors, which may offer more opportunities for part-time work (1).

Other appropriate responses will be accepted.

Hints and tips

What is the difference shown? Why might there be a difference between men and women in part-time work?

(ii) 6 marks (AO1 = 3 marks, AO2 = 3 marks)

This question examines the differences between gross weekly earnings across Great Britain. Your answer may focus on the importance of employment sectors and location. Some suggested ideas are given below but you may wish to expand on these or include other relevant points.

AO1 Demonstrating your knowledge and understanding

➤ There are significant differences in the amount of gross weekly earnings across Great Britain, with London and parts of the South East having the highest earnings, although some areas outside of London also have relatively high earnings.

➤ These differences can be linked to both location, e.g. proximity to London, and employment sector, e.g. primary, secondary, tertiary and quaternary.

AO2 Applying your knowledge and understanding

➤ Parts of London have very high gross weekly earnings because many people there are employed in high-income quaternary employment, such as financial and legal work.

➤ Parts of southern and southeastern England also have very high gross weekly earnings because people may be employed locally in high-income employment sectors, e.g. technology and defence.

➤ Deindustrialisation in some British regions may have caused job losses and median weekly wages to decline.

➤ Some more peripheral areas may have high incomes because of specialist economic activity, e.g. Cumbrian coast (nuclear industry) and Aberdeen (oil industry).

➤ Some peripheral regions may have people locally employed in the agricultural sector (e.g. parts of eastern England) or seasonal work (e.g. Cornwall), which tends to be lower paid.

Answers to this question will be given a mark within a level band

Level 1 (1–2 marks) You show limited knowledge about the difference in weekly earnings and demonstrate only a narrow range of understanding about the reasons for differences in weekly earnings in Great Britain. Part of your answer may be inaccurate or lack detail.

Level 2 (3–4 marks) You show mostly relevant knowledge about the difference in weekly earnings and demonstrate some understanding about the reasons for these differences within Great Britain. Some of your ideas may not be fully developed.

Level 3 (5–6 marks) You show accurate and relevant knowledge and understanding about reasons for differences in weekly earnings within Great Britain. Your ideas are detailed and developed fully, suggesting relevant examples linked to more specific locations.

Hints and tips

Where are the highest and lowest weekly earners? Is there a pattern that can be linked to employment sectors and opportunities? Are there some types of employment that attract higher wages than others? Can you include numbers to help explain your ideas?

(b) 6 marks (AO1 = 6 marks)

This question examines how people's quality of life can be affected by inequalities of pay. Relevant ideas are suggested below, but you do not need to include all of these in your response. Other relevant points will be credited.

AO1 Demonstrating your knowledge and understanding

➤ People who work in more skilled and professional employment receive higher pay than those in the primary sector and low-skilled services.

➤ People with seasonal and part-time work can receive lower pay than those with permanent contracts and full-time employment.

➤ Pay can be included in income levels. Income is a significant factor when assessing a person's quality of life as it can affect their ability to afford particular goods and services as well as being able to afford housing that satisfies their needs.

➤ Income can be used as part of composite indices that measure quality of life, such as the Index of Multiple Deprivation and Measuring National Well-being in the UK.

Answers to this question will be given a mark within a level band

Level 1 (1–2 marks) You show a basic knowledge of how people's quality of life can be affected by inequalities in pay and limited understanding of the impacts of pay on quality of life. Part of your answer may be inaccurate or lack detail.

Level 2 (3–4 marks) You show mostly relevant knowledge and understanding of the relationship between quality of life and pay inequality. Your answer may not be fully developed and may include only some reference to relevant examples.

Level 3 (5–6 marks) You show accurate and relevant knowledge and understanding of the relationship between quality of life and pay inequality. Your answer is detailed and uses a range of examples from both low- and high-income employment to show their effect on quality of life.

Hints and tips

Think about how income can affect quality of life in terms of access to basic needs, e.g. food, housing, economic security.

(c) 20 marks (AO1 = 5 marks, AO2 = 15 marks)

This question focuses on how different areas have different priorities, based on economic and social inequalities. This specification focuses on four different areas that will have different levels of need for regeneration (gated communities, 'sink estates', commuter villages and rural settlements). Suggested ideas are outlined below, but you do not need to include all of these in your response. Other relevant points will be credited.

AO1 Demonstrating your knowledge and understanding

➤ Regeneration relates to the need to reverse the economic and social decline of an area through new projects that may have a combination of both public money and private investment.

➤ Regeneration projects can help to change the built environment as well as the economic structure and social framework of an area.

➤ Parts of urban and rural areas may have differing regeneration needs and can have different levels of priority for regeneration.

AO2 Applying your knowledge and understanding

➤ Some more deprived areas, such as 'sink estates' in urban areas and declining rural settlements, can be in relative need of regeneration strategies.

➤ Some less deprived areas, such as gated communities in urban areas and commuter villages, are less likely to need regeneration.

➤ Economic and social differences in these places lead to different requirements for regeneration.

Answers to this question will be given a mark within a level band

Level 1 (1–5 marks) You include isolated points of knowledge and understanding of differences in economic and social inequalities and their connection with regeneration priorities, with some errors and inaccuracies. You do not make connections from the question to points made. Your answer is incoherent and lacks relevant evidence to support ideas. Your argument is limited, with unbalanced points. Points that you make are concluded in a general manner, if at all.

Level 2 (6–10 marks) You make some points showing knowledge and understanding of differences in economic and social inequalities and their connection with regeneration priorities, some of which may be relevant. You make some inaccurate points. You apply some of your knowledge about differences in economic and social inequalities and their connection with regeneration priorities, but your ideas are not developed or may not be linked directly to the question. You use some evidence to support statements, which may answer only part of the question. You make a conclusion but this is drawn from often unbalanced ideas.

Level 3 (11–15 marks) You make generally relevant points showing knowledge and understanding of differences in economic and social inequalities and their connection with regeneration priorities. Your ideas are mostly accurate and some connections are made between ideas. You interpret the question well in general but there may be some gaps in the use of evidence to support points. You draw a conclusion that links to the arguments made, but it is not fully supported by evidence.

Level 4 (16–20 marks) You show good use of knowledge and understanding of differences in economic and social inequalities and their connection with regeneration priorities. You make a range of relevant points to create a coherent argument supported by appropriate evidence. You apply your knowledge well throughout. All points you make are linked to the question. You draw a good, well-balanced conclusion that links clearly to the evidence presented.

Hints and tips

What type of places are in need of regeneration? Why is regeneration needed in these places? Think about how different places, such as 'sink estates', declining rural areas, gated communities and commuter villages, vary economically and socially.

Question 4 example responses

Student A

(a) (i) Figure 1 shows that in the UK in 2015 there were <u>nearly three times more women working part-time</u> than men. This could be because of increased flexibility in the workplace. <u>Flexible work opportunities</u> may be available in many of the professions which are more likely to have female employees, such as the care sector. Also, some women may choose to have flexible work patterns <u>after returning from maternity leave.</u>

ⓔ **This is a good answer, supported by data from the source. A valid reason is given, backed up with supporting statements.** **3 marks**

'nearly three times more women working part-time' Shows understanding of the figure in relation to the question.

'Flexible work opportunities' Reason given for difference with some expansion.

'after returning from maternity leave' Reason expanded further.

(ii) Figure 3 shows a wide variation in both the amount of money earned each week across Great Britain (between £391 and £958) and where these people work. In general, the <u>higher earners are found in southeast England</u>, although there are some anomalies to this general pattern. There are several reasons why people in some areas may earn the highest range (£600–£958 per week). For example, many of these people are found in <u>central and western London, where there are high concentrations</u> of high-income jobs, such as marketing, finance and legal occupations. This is because London is not only the national capital but also a global hub city. Workers may also get an extra '<u>London allowance</u>' for working here, which will further increase their wages. There are also other areas outside of London which attract high wages. This could be because of the <u>technical economy (e.g. Oxford</u> and Cambridge), defence (central southern England), and the energy sector (Cumbrian coast and northeast Scotland). However, there are also areas which have low levels of weekly earnings (between £391 and £449). This could be because more employment here is in traditionally lower income sectors (e.g. farm workers in Lincolnshire) or because the work <u>may be seasonal</u> (e.g. the tourism and hospitality sector in Devon and Cornwall). In rural areas, workers may also have less access to higher paid city-based jobs because the transport system is not good for regular commuting.

ⓔ **This is a good answer that suggests a range of reasons for the location of differences in wages in the time allowed. Both employment sectors and type of employment are included and regional examples linked to specific occupations are given. Good knowledge and good understanding are shown.** **Level 3, 6 marks**

'higher earners are found in southeast England' Understanding of pattern of main locational difference.

'central and western London, where there are high concentrations' Reason linked to location and employment type, with detail.

'London allowance' Reason for high wages in London expanded.

'technical economy (e.g. Oxford)' Reasons given for areas with high wages, with examples.

'may be seasonal' Reason for low wages expanded.

(b) In 2016, there were nearly 32 million people in paid employment in the UK, with some of those not in work relying on other forms of income such as pensions and universal credit. However, the difference in how much money people earn can affect their quality of life and life chances. The earning potential of a person depends on many factors, including their level of education, the sector in which they work and the type of employment, e.g. zero-hours contract or part-time. Some workers, such as those in some customer-service-orientated employment, are in the bottom 10% of earners. Even if they work full-time, they earn less than people in other professions and this will affect how much they have to spend on basic needs as well as their disposable income. Some low-income earners work full-time but may still rely on government help through benefits as well as food banks at certain times of the year. This adversely affects their quality of life as it decreases their economic security and may lower their self-esteem. It will also negatively affect the value of the Index of Multiple Deprivation for an area, indicating a lower quality of life. However, some professions such as those in the banking and finance sector have higher rates of pay. This may enable workers to afford better quality housing or housing in areas more convenient for work and be reflected in high values when measuring national wellbeing in the UK. However, as well as pay inequality, there are other aspects that can affect people's quality of life, such as health, education and training, happiness and the local and natural environment.

ℯ **The student has shown they understand the link between pay and quality of life. Examples from both low- and high-income earners are used to illustrate different aspects that could be included in quality of life analysis. The student shows that factors other than pay are important when considering quality of life.** **Level 3, 6 marks**

'nearly 32 million people in paid employment in the UK, with …' Sets the scene well in terms of the question.

'depends on many factors' Reasons given for differences in pay.

'spend on basic needs as well as their disposable income' Pay related to key needs.

'decreases their economic security and may lower their self-esteem' Low pay levels clearly linked to quality of life factors.

'Index of Multiple Deprivation' Answer extended to include contribution to a multivariate index.

'afford better quality housing or housing in areas more convenient' Comment about more wealthy earners and quality of life.

'other aspects' Good comment about other factors that can also influence quality of life.

(c) Regeneration strategies are schemes that are implemented to improve the quality of life of the people who live and work in an area. As these schemes often require a large amount of financial investment from public and private investors, priorities are set so that the areas which are most deprived can receive the most help. Regeneration schemes can focus on improving the economic situation of families to help them improve their standard of living. For example, sink estates are places which have clusters of poverty and crime, and often have higher percentages of people gaining help from the state through the benefit system or through schemes like free school meals. Social inequalities may be addressed through policies such as pupil premiums, which give extra money to schools to help disadvantaged children. For example, a primary school on the Broadwater Farm estate in north London receives above-average funding for disadvantaged children, including free school meals.

Declining rural settlements also have a higher priority for regeneration strategies, often because they are remote. While they may score highly in terms of environmental quality, there are often not enough people to support basic services, such as a grocery store or a primary school. For example, some villages in rural Wales have lower levels of mobile phone and broadband coverage than the rest of the UK. Upgrading technology here would have a large impact on businesses and local residents.

However, there are some areas which have a lower need for regeneration because they are wealthier and more connected. For example, the Itchen Valley in Hampshire has many villages that are well connected through road links such as the M3 and are close to Winchester, which has fast trains up to London. There is less need for regeneration here as there is enough population to maintain services such as a primary school. However, policies may be needed to address the issue of rising house prices so that local people are not priced out of the market.

In some urban areas, gated communities can enforce inequality in an area and may lead to changes in regeneration priorities for more deprived neighbouring communities. Gated communities are groups of properties that are highly protected. The streetscape may be designed to reduce crime and there are high levels of CCTV and other methods of surveillance. Higher income groups can also start to cluster together in areas that have been gentrified, such as parts of east London. Where once an area was prioritised for regeneration because of the concentration of lower income families, higher income earners are moving in and continuing the process of renewal, excluding long-time local residents from the process.

It is true that differences in economic and social wellbeing are considered when deciding which areas are in need of regeneration. More deprived areas, often in cities, are a higher priority when judging which places need additional help to progress. However, the economic and social situation of these areas can change by public policy or market forces and so priorities for regeneration areas change over time.

(e) **The student addresses the question well. A good knowledge of a range of places that have different regeneration priorities and a clear understanding of some economic and social needs are demonstrated. The student starts to evaluate the view by suggesting that there are some aspects of less deprived areas that may need government help in the future. A judgement is made in the conclusion. More specific case study information for the rural Wales example would improve the answer.** **Level 4, 17 marks**

Student B

(a) (i) The chart shows that <u>more women than men work part-time</u> in the UK. This may be because some women prefer to work part-time after having children as it may be <u>easier for them to work the family's childcare arrangements</u> around their job than their partner's.

ⓔ **The student has read the figure correctly and recognises that more women than men work part-time. However, while a reason is given, this is not expanded, e.g. are there some careers that people perceive as more 'flexible' than others?**

'more women than men work part-time' The key difference identified.

'easier for them to work the family's childcare arrangements' Reason for difference suggested.

(ii) There are places in Great Britain where people earn more money than others. Many people earn a lot of money in <u>London because they can find high-paid jobs in the quaternary</u> sector there. Lots of people who work in London have a degree or extra qualifications and this will help them to get a job which <u>pays better</u> (e.g. people who work in banks). However, there are many places in Great Britain where people do not earn as much. This is because they may live in the <u>countryside and many jobs there do not pay as well.</u> The <u>population density of some areas is lower</u> and so there are not the people to support higher paid jobs, e.g. parts of Wales.

ⓔ **The student shows a basic understanding of the general pattern of differences in wages and gives some general valid reasons for it. Good reasons for high wages in London are given, but for a balanced answer there should also be an attempt to give examples of employment types that may attract low wages. Specific place names and numbers from the resource would give detail to the answer.** **Level 2, 3 marks**

'London because they can find high-paid jobs in the quaternary' Reason suggested, with an example of location and employment type.

'pays better' Reason expanded.

'countryside and many jobs there do not pay as well' Examples of place and employment type are needed to gain a mark.

'population density of some areas is lower' Reason suggested with locational example given.

(b) Although many people work in the UK, they earn different amounts of money depending on what they do. Workers who have had many years of education and training, such as doctors and architects, are <u>likely to earn more</u> than those who have jobs requiring lower-level skills. If people <u>do not earn enough money to be able to afford</u> decent housing, food and other essentials, this can reduce their wellbeing. Also, the uncertainty of exactly how much they can earn per month can be stressful and this can affect them in a negative way. For example, some workers in the hospitality industry are on zero-hours contracts. This means that they only earn money when work is available. Although this means people have more flexibility about their working hours, they might earn less money than they need at the end of the month. This lack of economic security and income <u>reduces their quality of life.</u>

(e) The student shows some understanding of the link between differences in pay and quality of life and includes some relevant detail, e.g. zero-hours contracts. However, this could be expanded to include detail on how pay inequality can affect the quality of life of higher income earners, and income could be linked to wider measurements of wellbeing such as the Index of Multiple Deprivation. **Level 2, 3 marks**

'likely to earn more' Reason given for pay differences.

'do not earn enough money to be able to afford' Understanding shown of link between pay and quality of life.

'reduces their quality of life' Type of employment linked to impact on quality of life.

(c) These areas have many problems that can affect the standard of living for people who live there. There are some inner-city areas that have high crime rates and people who are struggling to get by. The government can help people by raising the minimum wage, giving Healthy Start vouchers for children in low-income families and building more affordable houses. Some people find it difficult to do the basic things in life such as pay rent for their home. These regeneration strategies will help people to improve their quality of life and this will help the area to progress. Regeneration can also include building new facilities for people, such as a pharmacy or a nursery. There are also some rural places that also suffer from deprivation. Wages are low and the jobs are only there in some parts of the year because they are linked to tourism and farming. People move out to cities to find better paid jobs. This means that local businesses lose their customers and they may have to shut down. It takes more time to get into larger towns because public transport is not good. There also may be poor internet coverage so it may take ages to download the site that you want. However, there are some less deprived places which do not need regeneration strategies. People here generally have higher incomes and the services that they need are close by. The government does not need to give extra money to these places to help them and they should spend the money in places that need it most. However, people can view places differently meaning that where one group might think regeneration is a high priority, another might not agree. Also people may disagree about what type of regeneration should take place. There is only so much money to go around and so these decisions are difficult. I agree with the view that places which are more deprived should be prioritised for regeneration schemes.

(e) The student demonstrates some knowledge that different areas have different regeneration needs based on economic and social inequalities. There is some understanding that some places are more deprived than others, but ideas are not clearly linked to specific examples, e.g. the 'inner city' is too general — the focus should be on 'sink estates', allowing for more detail and better development of points made. There is some knowledge about specific government schemes used to help social deprivation, and some knowledge and understanding of less deprived places; again, place-specific detail is lacking. The conclusion is very brief and lacks balance. **Level 3, 12 marks**

Area 3 Physical systems and sustainability

Topic 5 The water cycle and water insecurity

Topic 6 The carbon cycle and energy security

Question 1 mark scheme

(a) 3 marks (AO1 = 2 marks, AO2 = 1 mark)

You gain 1 mark for analysing the graph and up to 2 marks for explanation.

➤ From January to April (A), there is a water surplus.
➤ This means that precipitation is greater than evapotranspiration, and the soil water store keeps being recharged.
➤ From May to July (B), evapotranspiration exceeds precipitation — but there is still soil moisture available, as the moisture in the stores can be accessed by plants.
➤ However, by late July and to the end of September (D), the soil moisture store has been used up and so there is water deficiency.

Hints and tips

Show clear understanding of the processes that cause the changes in soil moisture content in your answer. Name them, and show that you know the role they play.

(b) 6 marks (AO1 = 6 marks)

Some suggested ideas are given below but you may wish to expand on these or include other relevant points.

AO1 Demonstrating your knowledge and understanding

Various physical factors are relevant:

➤ Climate: Influences amount of precipitation input and evaporation output; also influences vegetation type, which in turn influences interception flows.
➤ Soils: Influence the rates of infiltration, runoff and throughflow flows.
➤ Geology: Influences percolation and groundwater flows.
➤ Relief: Influences runoff output and precipitation.
➤ Vegetation: Influences interception, infiltration, runoff and amounts of transpiration output.

Answers to this question will be given a mark within a level band

Level 1 (1–2 marks) You show limited geographical knowledge and a narrow understanding of the physical factors in a drainage basin. Part of your answer may be inaccurate or lack detail.

Level 2 (3–4 marks) You show mostly relevant geographical knowledge and understanding of the physical factors in a drainage basin. Some parts of your answer are not fully developed.

Level 3 (5–6 marks) You show accurate and relevant geographical knowledge and understanding of the physical factors in a drainage basin. Your answer is detailed and fully developed.

Hints and tips

Identify a range of factors, and link them clearly to the processes at work in the drainage basin.

(c) 8 marks (AO1 = 8 marks)

Some suggested ideas are given below but you may wish to expand on these or include other relevant points.

AO1 Demonstrating your knowledge and understanding

Physical factors

Meteorological drought:
➤ Short term: For example falls in precipitation levels.
➤ Longer term: Underlying trends in precipitation, e.g. El Niño Southern Oscillation (ENSO).
➤ Compounded by other meteorological factors such as higher temperatures/stronger winds producing more evaporation output.

Hydrological drought
➤ Decreased precipitation inputs result in reduction in stream flows, reservoirs and aquifer levels.
➤ Increased temperatures resulting in salinisation and reduction in water quality.

Human causes
➤ Population growth increases pressure on the land and decreases carrying capacity.
➤ Overgrazing and overcultivation and deforestation can reduce evapotranspiration and lower rainfall levels.

Human-induced climate change could be a factor increasing the frequency or duration of drought.

Answers to this question will be given a mark within a level band

Level 1 (1–2 marks) You show limited geographical knowledge and a narrow understanding of the physical and human factors. Part of your answer may be inaccurate or lack detail.

Level 2 (3–5 marks) You show mostly relevant geographical knowledge and understanding of the physical and human factors. Some parts of your answer are not fully developed.

Level 3 (6–8 marks) You show accurate and relevant geographical knowledge and understanding of the physical and human factors. Your answer is detailed and fully developed.

Hints and tips

Show good depth of understanding of both factors through the use of examples to illustrate your points.

(d) 12 marks (AO1 = 3 marks, AO2 = 9 marks)

Some suggested ideas are given below but you may wish to expand on these or include other relevant points.

AO1 Demonstrating your knowledge and understanding

Energy security refers to the uninterrupted availability of energy sources at affordable prices and includes aspects such as availability/accessibility, affordability and reliability.

➤ The graph shows an increase over time of energy consumption, initially slowly up to the 1950s, then very rapidly.

➤ A small proportion of this energy mix is from recyclable (nuclear) and renewable (hydro) sources.

➤ The majority is from fossil fuels.

Implications for energy security include:

➤ Demand.

➤ Availability: Are the resources available in a particular country? If not, they will incur transport costs, which could drive consumption of this particular source down.

➤ Accessibility: How accessible are the energy sources that are available? Exploiting them may be costly.

➤ Economic development: There is a strong correlation between level of development and energy consumption, as the technology that improves standards of living for more developed countries also drives up energy demand.

AO2 Applying your knowledge and understanding

Link between supply and demand of fossil fuels:

Coal

➤ The consumption of coal is declining compared with oil and natural gas, but its production is still increasing, especially in China and the USA.

➤ There is a close correlation between the countries that consume and those that produce coal, reflecting the cost of transporting coal.

Oil

➤ Nearly half the world's supply comes from two major groups/regions: the Organization of the Petroleum Exporting Countries (OPEC) and North America.

➤ Europe, one of the biggest consumers, is not a major producer.

Gas

➤ Global production is dominated by the USA and Russia.

➤ The top five gas importers, including Germany and the UK, are not major producers.

Challenges of renewable energy sources:

➤ Not all countries can exploit renewable sources, e.g. they may have no coasts, not enough sunshine, no geothermal rocks.

➤ Only those countries with good hydro renewable sources are likely to be able to replace fossil fuels.

➤ Renewables can have detrimental environmental impacts, e.g. hydroelectric power (HEP) reservoirs flood valleys; wind farms are objected to on visual pollution grounds.

➤ Nuclear power could potentially meet increasing demands, and it is a recyclable source. However, there are issues related to it including safety, security (terrorism threat), disposing of radioactive waste.

Answers to this question will be given a mark within a level band

Level 1 (1–4 marks) You show only a limited geographical knowledge and understanding of the implications of the changing consumption patterns. You make limited connections between aspects of your answer and support your interpretations with limited evidence. You draw unbalanced conclusions based on the material in your answer.

Level 2 (5–8 marks) You show mostly relevant and accurate geographical knowledge and understanding of the implications of the changing consumption patterns. You make mostly relevant connections between aspects of your answer as appropriate and support your interpretations with some evidence. You draw conclusions based on the material in your answer but your conclusions may be limited or unbalanced.

Level 3 (9–12 marks) You show relevant and accurate geographical knowledge and understanding of the implications of the changing consumption patterns. You make sound connections between aspects of your answer as appropriate and support your interpretations logically with evidence. You draw balanced and logical conclusions based on the material in your answer.

Hints and tips

Remember to weigh up the various elements relevant here to come to an overall assessment at the end, based on the argument you have presented.

(e) 20 marks (AO1 = 5 marks, AO2 = 15 marks)

Some suggested ideas are given below but you may wish to expand on these or include other relevant points.

AO1 Demonstrating your knowledge and understanding

There are three main threats to the biological carbon cycle and water cycle:
- Climate change
- Growing demand for food
- Ocean acidification

AO2 Applying your knowledge and understanding

Climate change
- Additional carbon dioxide in the atmosphere enhances the greenhouse effect, raising temperatures — this produces more water vapour in the atmosphere, acting as a feedback loop reinforcing warming and affecting both stores.
- Climate belts are shifting because of climate change. For example, rainfall totals are falling in the Amazon Basin, resulting in more drought. This also affects the carbon cycle as the forests, during periods of drought, can become net emitters of carbon dioxide.
- Climate change can affect the carbon cycle positively, in that more carbon dioxide can promote more plant growth. However, there appears to be an upper threshold to the relationship between carbon dioxide and plant growth, as plants need access to other things such as water and nitrogen to grow.

Growing demand for food
This results in land conversion, changing land use from the natural ecosystem use. Land conversion includes:
- deforestation
- afforestation
- grassland conversion.

Land conversion has impacts on the biological carbon and water cycles.

Ocean acidification
- Increased carbon dioxide emission has resulted in an increase in carbon dioxide in the oceans, leading to a significant fall in the pH values.
- This can have negative impacts on coral reefs and associated ecosystems.

Area 3 Physical systems and sustainability

Answers to this question will be given a mark within a level band

Level 1 (1–5 marks) You include isolated points of geographical knowledge and understanding of the extent to which climate change is a significant factor, with some errors and inaccuracies. You have not made connections from the question to points made. Your answer is incoherent and lacks relevant evidence to support ideas. Your argument is limited, with unbalanced points. Points that you make are concluded in a general manner, if at all.

Level 2 (6–10 marks) You make some points showing geographical knowledge and understanding of the extent to which climate change is a significant factor, some of which may be relevant. You make some inaccurate points. You apply some of your knowledge but your ideas are not developed or may not be linked directly to the question. You use some evidence to support statements, which may answer only part of the question. You make a conclusion but this is drawn from often unbalanced ideas.

Level 3 (11–15 marks) You make generally relevant points showing geographical knowledge and understanding of the extent to which climate change is a significant factor. Your ideas are mostly accurate and some connections are made between ideas. You interpret the question well in general but there may be some gaps in the use of evidence to support points. You draw a conclusion that links to the arguments made but is not fully supported by evidence.

Level 4 (16–20 marks) You show good use of geographical knowledge and understanding of the extent to which climate change is a significant factor. You make a range of relevant points to create a coherent argument supported by appropriate evidence. You apply your knowledge well throughout. All points you make are linked to the question. You draw a good, well-balanced conclusion that links clearly to the evidence presented.

Hints and tips

Consider the effect of climate change, both positive and negative, and come to an overall conclusion based on the argument you outline.

Question 1 example responses

Student A

(a) The levels of soil moisture vary throughout the year as the inputs and outputs vary. At A (January to April), there is a soil moisture surplus. This is because the inputs from precipitation are greater than the outputs from evapotranspiration and the soil moisture store is kept topped up.

As we move into the summer and temperatures rise, so rates of evapotranspiration increase. However, at B (May to early July), plants can still access water as they use the water stored during A.

However, towards the end of the summer at D (late July to September), evapotranspiration is still greater than rainfall, and crucially the soil moisture store has been fully utilised, resulting in the soil moisture deficiency shown in the graph at that time.

ⓔ **The student analyses the figure well to pick out the changing patterns over the months, and they use their understanding effectively to explain the patterns in detail.** **3 marks**

(b) Climate can have an impact on precipitation inputs in a drainage basin and evaporation out of it. For example, in many tropical areas rainfall tends to be seasonal. The Blue Nile experiences next to no rainfall for much of the year, but it has a rainfall peak during the summer. In contrast, the rainfall in the UK is reasonably constant throughout the year. However, river levels still vary here, as during the summer the warmer temperatures produce more evaporation so there is less water in the drainage basin overall.

Second, soils can affect flows in the drainage basin. Where you have clay soils, less infiltration can occur as the soils are more impermeable. This has the effect of reducing throughflow and instead creating more surface runoff. Where the soils are more sandy, the opposite is the case — the more permeable soils allow for more infiltration, increasing sub-surface flows (throughflow) and reducing surface runoff.

Under the soils, we have our third factor: geology. Where rocks are more permeable, they will allow soil moisture to percolate deeper and increase groundwater flows. This also allows underground aquifers to develop, producing an important water resource in many of the world's most arid regions.

The topography of the landscape can also have an impact. Where you have steeply sloping areas, precipitation is more likely to experience surface runoff due to the influence of gravity. In addition, orographic uplift over mountains can increase precipitation totals there. These two elements combine to produce some of the major flooding events in places like Cumbria. In addition, altitude can mean that some of the precipitation falls as snow and is stored in snowpacks on mountains. This can produce runoff and lead to flooding in the following spring as the snow melts.

(e) **This answer covers a broad range of geographical ideas linking physical factors to inputs, flows and outputs and develops them in detail.** **Level 3, 6 marks**

(c) Drought is defined as a deficiency of water over an extended time period (normally at least one season) and it occurs as a result of a complex interplay of physical and human factors.

There are two main physical factors to consider. First, meteorological factors. Precipitation levels can vary over the short term. Perhaps one year has below-average values of rainfall which can reduce water supplies in the drainage basin and cause drought. This can be compounded by some other meteorological factors that can accompany lower rainfall totals. Especially in arid and semi-arid areas, there are often stronger winds and higher temperatures when drought is occurring. These can combine to increase outputs of water via evaporation, increasing the drought risk. There can also be longer-term meteorological falls in precipitation totals. For example, in the Sahel region of Africa, rainfall totals have been generally below the 30-year average since the 1970s, with some years such as 1985 being nearly 40 mm below average. Wider-scale meteorological factors can also be relevant. For example, the ENSO can lead to changes in the distribution of warmer water in the Pacific. During El Niño years, the warmer water stays further to the east of the ocean. As a result, there is an increased drought risk in Australia and parts of Southeast Asia.

The second physical factor is hydrological. When there are reductions of inputs of water into a drainage basin, this can affect stores and transfers in the basin, contributing to drought. Stream flows can be reduced, and this can create problems if people rely on these for their water supply for drinking or irrigation. Reservoir levels may drop, and also the volume of water in underground aquifers, as they are not being recharged. In addition, evaporation of water can result in salinisation of the ground and a reduction in water quality in the rivers.

Drought can occur naturally, but various human actions can contribute to the drought risk too. They can act as a positive feedback loop, magnifying the effect of physical factors. The Sahel region in Africa illustrates this. There have been a number of serious drought events there, such as the 1999–2000 drought, during which 10 million people needed food assistance. As mentioned above, drought here is triggered by meteorological factors. But various human factors exacerbate the drought risk here. The region is marked by rapid rates of population growth which puts more demand on the natural environment, leading to environmental degradation. For example, overgrazing and overcultivation mean that the soil is no longer protected and is more vulnerable to erosion. Any rain that does fall is less likely to be retained by the soil. Removal of trees for firewood also increases the risk of soil erosion. As agriculture here is supported by rainwater, any falls in rainfall totals can lead to significant impacts.

(e) **The student shows accurate and relevant geographical knowledge throughout. A wide range of ideas are covered and generally explained in detail. The human factors could be linked more clearly to the drought, but it is still well handled overall.** **Level 3, 7 marks**

(d) Energy security refers to the uninterrupted availability of energy sources at affordable prices and includes aspects such as availability, accessibility, affordability and reliability.

Underlying the challenges to global energy security are increases in demand. The graph shows that, prior to the 1950s, consumption was increasing, but at a relatively low rate, only reaching around 100 exajoules/year by that date. Since then, consumption has risen fast, reaching around 550 exajoules/ year only 60 years later in 2010. The graph also indicates that only a small proportion of this increase in energy use is from renewable and recyclable sources — the majority of it is coal and oil, followed by natural gas.

This growth is projected to increase, with energy demand globally expected to grow from 11 billion tonnes of oil equivalent in 2005 to nearly 18 billion by 2030.

The patterns in the link between supply and demand of fossil fuels are interesting. When it comes to coal, overall consumption rates are in decline compared with oil and natural gas. However, production is still increasing, especially in China and India. There is a close correlation between those countries that produce coal and those that consume it, reflecting the costs involved in transporting coal. Oil and natural gas are slightly different. In both cases, production is more geographically concentrated (nearly half the world's oil comes from OPEC countries and North America; global production of natural gas is dominated by the USA and Russia).

These trends have a series of significant impacts for energy security. For instance, there are questions about energy availability. Although most major energy producers are also energy consumers, many of the major consumers are not producers as they may not have fossil fuel reserves in their countries. This adds to costs of importing the fuel, driving up prices and potentially raising issues of affordability for some people within the country. In the future, it is likely that the USA will loosen some of the restrictions that were placed on coal production to increase availability of home-grown energy, increasing energy security.

There are also questions about accessibility. If energy sources are deep underground and inaccessible, then that can raise the cost of extraction and affect the affordability of the energy produced. Some technological solutions to this (such as fracking) are controversial. Shale gas extraction in the USA not only supports 600,000 jobs, but it also provides affordable gas for chemical, manufacturing and steel industries there. In 2000, shale gas made up 1% of the USA's gas supply — by 2015 this had risen to 15%. However, many believe there to be serious environmental issues related to fracking, including methane emissions, high water consumption, water contamination, earthquake risks and health issues.

In response to the issues with fossil fuels, there is an increasing global emphasis on the development of renewable and recyclable energy sources. However, there are issues with these sources in terms of energy security.

In terms of availability, not all countries have readily exploitable renewable sources at hand. Some countries do not have coastlines so tidal energy is not available. Many parts of the world do not have sufficient hours and intensity of sunlight to make solar energy viable on a large enough scale. Many countries cannot access geothermal energy in the way countries like Iceland can.

In terms of reliability, therefore, for most countries many of the renewable sources would struggle to meet demand. Only those countries with access to enough hydro renewable sources (e.g. HEP) are likely to be able to meet energy demands. However, even with these sources, there are controversies. For instance, the environmental impacts of flooding resulting from HEP reservoirs are considerable. There is also often a public outcry at the siting of wind farms. In fact, to ensure a reliable energy supply, many countries are factoring nuclear power into their energy mix. It could potentially meet increasing demands, and it is a recyclable source. However, there are issues related to it including safety (either from malfunction, e.g. the Chernobyl incident, or natural disaster, e.g. the Fukushima nuclear power plant following the 2011 Japanese tsunami), security (terrorism threat), disposing of radioactive waste, etc.

Overall, the increasing demands for energy consumption place significant pressures on energy security across the globe. While there are various ways of trying to meet demand and achieve energy security, each of them faces various issues and challenges and none is a panacea.

ⓔ **The student applies their knowledge well to extract relevant trends and patterns from the graph. The understanding shown throughout is detailed and relevant. Connections are made between various aspects of the issue. The student assesses the aspects well and produces a balanced and coherent argument.** **Level 3, 11 marks**

(e) Across the world, the biological carbon and water cycles are being threatened by various factors, including climate change caused by human activity.

Climate change firstly is affecting the carbon and water cycles because of the additional release of carbon dioxide into the atmosphere from the burning of fossil fuels. About 45% the carbon dioxide released into the atmosphere since the beginning of the industrial revolution remains there. This enhances the greenhouse effect by altering both the carbon and water cycles. The carbon cycle is altered by the fact that the additional carbon dioxide that has been released is now stored in the atmosphere. But this has also set up a feedback loop which has affected the water cycle. Carbon dioxide itself actually only contributes about 20% of the Earth's greenhouse effect; water vapour contributes 50%. But the amount of water vapour in the atmosphere is influenced by the amount of carbon dioxide. Carbon dioxide remains a gas at a wider range of temperatures than water vapour. So increased levels of carbon dioxide act as the kick-start to warm the atmosphere, which produces more water vapour. So, both cycles are impacted.

Furthermore, climate change is affecting the cycles owing to the shifting in climate belts that is accompanying our warming planet. The Amazon tropical rainforest provides an illustration of this. The rainforest is a vital carbon sink, absorbing around 2 billion tonnes of carbon dioxide per year. This ecosystem also interacts with the atmosphere, providing feedback loops that keep precipitation levels higher here. The Amazon River may discharge 17 billion tonnes of water into the ocean every day — but the forest produces 20 billion tonnes of water vapour into the atmosphere above it each day. This humidity lowers the pressure here, drawing in moist air from the Atlantic Ocean. However, since 1990, drought has been more common in the rainforest regions. In 2005 and 2010, drought in the Amazon Basin turned the forests from carbon sinks to net contributors of carbon, releasing around 5 billion tonnes for each event.

On the other hand, climate change can be seen to have some positive impacts on the carbon cycle. More carbon dioxide in the atmosphere can lead to an increase in plant growth. A study in 2016 suggested that, across the globe, there has been an increase of between 25% and 50% of the vegetated land across the planet. This may act to mitigate further global warming as this increased vegetation absorbs carbon dioxide. However, others are sceptical of this claim, as plant growth relies on other factors also, including water and nitrogen supplies. They claim that, if access to these is limited, the plants will reach a threshold beyond which more carbon dioxide in the atmosphere will not result in more plant growth.

We can see that climate change has some significant impacts on the carbon and water cycles, but other factors affect them also. For example, increased global demand for food. This can lead to land conversion, which is the change from the natural ecosystem to an alternative land use. One land use change is deforestation. By 2015, around 30% the world's forests are estimated to have been lost. About half of this is due to commodity production (including food); other reasons include open-cast mining, dams and reservoirs, and infrastructure developments. The loss of forest impacts the carbon cycle in various ways: it reduces the amount of carbon dioxide stored in the carbon sink and it reduces the intake of carbon dioxide via photosynthesis, leaving more of the gas in the atmosphere. In addition, when the forest is cut down, it is often burned to clear it. This releases more carbon dioxide directly into the atmosphere. The loss of forest can also affect the water cycle. There is less interception of water and less infiltration — this can reduce the water in the groundwater store. Less interception also means that there is less evaporation off the leaves and this, coupled with reductions in transpiration, means that the rainfall totals can fall in places where deforestation occurs and downwind from these places too. In Brazil, São Paulo has suffered a water crisis linked to deforestation to the west.

A third factor affecting the cycles is ocean acidification. This is resulting from the increased burning of fossil fuels releasing carbon dioxide into the atmosphere. About 30% of the carbon dioxide released into the atmosphere since the beginning of the industrial revolution has been absorbed by the oceans. This has caused the pH of the oceans to fall from 8.2 to 8.1 since 1750 — a fall of 30%. As the pH drops, coral reefs can be affected. Coral stops growing below pH 7.8 (a level we could reach by 2100), so ocean pH could cross a threshold resulting in permanent damage to the coral reefs and the ecosystem dependent on them.

To conclude, climate change caused by human activity has serious implications for both the biological carbon cycle and water cycle. It could be argued that the water cycle is being affected directly now, because climate change is disrupting climate belts and weather patterns which in turn changes levels of precipitation. This affects humans in terms of water shortages and flood disasters. Humans are directly reducing biological carbon stores through deforestation and land use change, but this makes climate change worse by releasing more carbon into the atmosphere — and long-term climate change will degrade more of the world's forests and other biomes.

(e) **This is a detailed answer which shows accurate knowledge and understanding throughout. Interpretations are backed with evidence and examples. The student explores connections well and produces a well-balanced argument and reaches a logical conclusion. To further improve marks, a more balanced approach to dealing with the three main causes is needed.**

Level 4, 18 marks

Student B

(a) Initially, there is a water surplus at A (January to April), followed by a period of time when the soil moisture is being used up at B (May to early July) and then a period of soil moisture deficiency (late July to September).

℮ **The patterns shown in the graph are outlined.**

This is because the soil moisture has gradually been used up as the year goes on and eventually there is not enough moisture in the soil, resulting in the deficit.

℮ **The explanation of the changing patterns, although broadly correct, lacks detail and misses out some key elements.** **2 marks**

(b) Soils affect drainage basin processes. They can affect infiltration — if the soil is impermeable, then there will be less of this occurring. As a result, the amount of throughflow increases and runoff decreases. The opposite is the case with permeable soils.

℮ **This point about throughflow is incorrect.**

Vegetation is another factor. More trees means more transpiration from the leaves. Leaves catch the water as it falls and this can affect how it moves through the drainage basin.

℮ **This point is underdeveloped and lacking detail.**

In steep areas, you are going to get more water running down the sides of the hills. You will probably also get more rain on hilltops too.

Some areas get more rain overall, and some get less. If you live in an area that gets more rain, then there will be more water in the drainage basins there.

℮ **These two points are also underdeveloped and lacking in detail. The student should mention the factors they are exploring as well.**

Urbanisation affects things too. If an area has lots of buildings and roads, this creates impermeable surfaces. So there will be less infiltration.

℮ **This point is irrelevant — it discusses human factors when the question asks for physical. Overall, this answer shows only isolated elements of geographical understanding. The points made are not consistently relevant, and they are underdeveloped and lacking detail. Level 1, 2 marks**

(c) Drought occurs when there is a decrease in rainfall totals. This can be due to changes in the amount of precipitation occurring if we have a year when there is less rain falling. If temperatures are higher and we have a warm summer, this can cause more evaporation of water which can cause droughts to occur. It might also be due to longer-term changes in rainfall patterns. In the Sahel, the rainfall totals have been much lower since the 1970s and this has resulted in some severe droughts there including in 1985. The ENSO in the Pacific can change rainfall patterns too. In an El Niño year, there can be drought in Australia as the rain falls further east in the Pacific Ocean.

Less water in the drainage basin can also result in drought. Less water flowing in rivers means there is less water available to people, especially in some poorer countries where they are more dependent on getting water from rivers. If aquifers dry up, this can leave people without a water supply as well.

Human actions can also cause drought. Again in the Sahel, there have been various human actions that have caused drought. Overgrazing and deforestation have degraded the environment, making drought more common. This has happened because of the rapid increases in population that have occurred here over the past few decades putting pressure on the environment.

ⓔ **A range of factors, both physical and human, is covered but points remain mostly underdeveloped. To gain more marks, the student should express their points more clearly to make sure the detail of understanding comes across.** **Level 2, 4 marks**

(d) The graph shows that consumption was increasing before the 1950s, but only slowly. Since the 1950s, consumption has risen quickly. Most of this growth was in fossil fuels, and very little renewable or recyclable energy. The graph shows that growth is probably going to increase in the future at the same fast rate.

This rapid increase in consumption will have impacts for energy security. There could be problems for energy availability. The countries that produce most coal, oil and natural gas tend to be the countries that consume much of it. However, for countries that are big consumers but not major producers and so have to import their energy source, this could cause prices to go up due to transport costs.

Energy security can also be affected by issues of accessibility. If energy sources are less accessible, technology such as fracking can help extract them. This can produce natural gas at a relatively cheap price. But some people have concerns about the issues around fracking, including water consumption, water contamination and earthquakes. It remains controversial.

So, renewable energy may help with availability. But not every country has available renewable sources. For example, there may not be enough sunlight for solar energy. Or a country may be in the middle of a continent and so not be able to use wave power. This means that renewable energy will probably not be reliable enough and have enough availability to meet energy demands. This means that countries are factoring in nuclear power to their energy mix. This is available, accessible and reliable and it is also recyclable energy. However, it is still controversial. How is the contaminated nuclear waste dealt with? What about safety issues, either from breakdown, natural disaster (such as the Japanese tsunami and Fukushima nuclear plant) or terrorist attack?

In conclusion, the increasing demands for energy across the globe raise many challenges for energy security. There are various possible solutions but these have their own issues, so the challenges remain significant.

ⓔ **The student makes a number of good and relevant points to answer the question. To gain more marks, they should extract more details from the graph, quoting figures to give more evidence to support their answer. Points should be developed in more detail, drawing on examples or deeper understanding to give more substance to them. An attempt is made to make judgements about the factors, but this should be done more fully throughout.** **Level 2, 6 marks**

(e) Climate change has had impacts on the water cycle and carbon cycle. Climate change is causing climate belts to shift. The Amazon Basin in South America is experiencing more droughts, like in 2005 and 2010. This affects the water cycle as the Amazon forest returns a lot of water into the atmosphere as water vapour. But when drought occurs, it also affects the carbon cycle. Usually, the forests act as a store of carbon dioxide. But, during those drought years, the forests gave off more carbon dioxide than they took in (around 5 billion tonnes). Climate change is also affecting the cycles in the tundra. The climate belts here are shifting north and the summers are getting warmer. This is producing more rain as part of the water cycle. But the increased temperatures are also increasing evaporation and so overall water supplies are projected to fall in this climate zone. This will lead to an increased fire risk with areas burned in fires expected to double in Alaska by the middle of the twenty-first century. This will add more carbon dioxide to the atmosphere, affecting the carbon cycle too.

The cycles in the atmosphere are also affected. The extra carbon dioxide in the atmosphere traps more heat in via the greenhouse effect. Warmer temperatures increase evaporation and produce more water vapour in the atmosphere. Water vapour is a more effective greenhouse gas and that creates more warming, leading to a feedback loop that impacts both cycles.

However, other factors affect the carbon and water cycles too. The increased amount of carbon dioxide in the atmosphere since the industrial revolution has led to more carbon dioxide in the oceans, making them more acidic. As this happens, corals are less able to absorb the calcium carbonate they need to maintain their skeletons and the reefs start to dissolve.

Land use can also impact the cycles. As demand for food increases across the world, more and more land is being given over to growing crops. This has resulted in high deforestation rates across the world — about 30% of the world's forests had been lost by 2015. The forests act as a vital store of carbon in the carbon cycle and less carbon can be stored when they are cut down. As well as that, trees are often burned when they have been cut down. This adds more carbon dioxide into the atmosphere. When the trees are cut down and burned, there is obviously less transpiration. But there is also less evaporation (the leaves of the trees would have intercepted the water and it could have evaporated from there). This changes the water cycle. On the other hand, there is a global attempt to plant more trees around our planet. This is called afforestation. This may help to offset these problems. But planting trees can have problems too — such as taking more water in to help them to grow.

Another land use change is grassland conversion. Grasslands are found in temperate regions such as the prairies and the savannah tropical regions. Both cycles are affected here when ploughing takes place. Soil degradation can reduce soil stores of carbon.

In conclusion, we can see that a variety of factors affect the carbon cycle and water cycle, not just climate change.

ⓔ **This answer is generally relevant and covers the elements needed to answer this question. However, there are a number of things that could be improved to score further marks. First, there is limited use of evidence to support points. More examples and figures would help. Second, although a conclusion is attempted, it is not as well supported by the preceding argument as it could be. More understanding of the various roles of the different factors needs to be shown throughout.**
                                                                                        Level 3, 11 marks

Question 2 mark scheme

(a) 3 marks (AO1 = 2 marks, AO2 = 1 mark)

You gain 1 mark for analysing the graph and showing how the two transfers change over time:
➤ Infiltration rates start high then decline, whereas surface runoff starts low then rises with time.

You gain 2 marks for explaining the changes:
➤ Initially the ground is dry, so there is a large capacity for infiltration.
➤ Over time, infiltration rates fall as the soil store becomes saturated with water.
➤ As infiltration is occurring at its maximum rate, any additional rain that falls is more likely to travel by surface runoff.

Hints and tips

Outline the changes and make a clear connection between them.

....

(b) 6 marks (AO1 = 6 marks)

AO1 Demonstrating your knowledge and understanding

Storm hydrographs show how a river's discharge varies in response to a rainstorm.

➤ At first, discharge remains largely unchanged, before rising quickly to a peak, then falling back to its original level.

➤ The response can be flashy (short lag time and a higher peak discharge) or flat (longer lag time and lower peak discharge) depending on a range of physical and human factors.

➤ Physical factors include basin size, relief, shape, soil type and geology.

➤ Human factors include urbanisation, agriculture and afforestation.

Answers to this question will be given a mark within a level band

Level 1 (1–2 marks) You show limited geographical knowledge and a narrow understanding of the physical and human factors. Part of your answer may be inaccurate or lack detail.

Level 2 (3–4 marks) You show mostly relevant geographical knowledge and understanding of the physical and human factors. Some parts of your answer are not fully developed.

Level 3 (5–6 marks) You show accurate and relevant geographical knowledge and understanding of the physical and human factors. Your answer is detailed and fully developed.

Hints and tips

Identify a range of physical and human factors and link them clearly to the processes at work in the drainage basin.

....

(c) 8 marks (AO1 = 8 marks)

AO1 Demonstrating your knowledge and understanding

Physical factors

➤ Rainfall: Intense storms can lead to flash flooding as the large amounts of rainfall exceed infiltration capacity; seasonal variations in rainfall, for example the Indian monsoon, can produce large quantities of rain during parts of the year.

➤ Snowmelt: Snow acts as a seasonal store of precipitation, allowing it to build up. When spring arrives and temperatures rise, this releases large amounts of stored water which can lead to flooding.

Human factors

➤ Land-use change: Deforestation reduces interception and evapotranspiration, resulting in more water in the drainage basin and more surface runoff of that water, increasing the flood risk.

➤ Hard engineering strategies can fail, increasing the magnitude of the flood event.

Answers to this question will be given a mark within a level band

Level 1 (1–2 marks) You show limited geographical knowledge and a narrow understanding of the physical and human factors. Part of your answer may be inaccurate or lack detail.

Level 2 (3–5 marks) You show mostly relevant geographical knowledge and understanding of the physical and human factors. Some parts of your answer are not fully developed.

Level 3 (6–8 marks) You show accurate and relevant geographical knowledge and understanding of the physical and human factors. Your answer is detailed and fully developed.

Hints and tips

Show good depth of understanding of both factors through the use of examples to illustrate your points.

(d) 12 marks (AO1 = 3 marks, AO2 = 9 marks)

AO1 Demonstrating your knowledge and understanding

➤ Volcanic outgassing is one of the processes whereby geological carbon is released into the atmosphere.

➤ Outgassing occurs at: subduction zones and ocean ridges; hot spots with no current volcanic activity such as Yellowstone Park; fractures in the lithosphere.

AO2 Applying your knowledge and understanding

When this carbon is released, a negative feedback loop can occur.

➤ In the short term, the extra carbon in the atmosphere results in an enhancement of the greenhouse effect, resulting in rising temperatures.

➤ This extra atmospheric energy causes more global precipitation.

➤ This in turn causes more chemical weathering and erosion of rocks, depositing ions on ocean floors, where it is absorbed into the geological store once more.

➤ This whole cycle takes hundreds of thousands of years.

➤ There are other processes that balance the geological carbon cycle on long timescales, such as carbon sequestration in the oceans by biological organisms that eventually locks carbon away as carbonate rocks (limestone) and burial of plant matter on land that locks carbon up as peat, coal and other fossil fuels.

➤ This means volcanic carbon emissions are only one part of the larger, long-term geological carbon cycle.

Therefore, we can see that, although volcanic outgassing causes short-term variations in climate through the release of carbon into the atmosphere, in the longer term, because of the negative feedback loop, it ultimately balances out this emission by contributing to the geological sequestering of carbon.

Answers to this question will be given a mark within a level band

Level 1 (1–4 marks) You show only a limited geographical knowledge and understanding of the role played by volcanic outgassing. You make limited connections between aspects of your answer and support your interpretations with limited evidence. You draw unbalanced conclusions based on the material in your answer.

Level 2 (5–8 marks) You show mostly relevant and accurate geographical knowledge and understanding of the role played by volcanic outgassing. You make mostly relevant connections between aspects of your answer as appropriate and support your interpretations with some evidence. You draw conclusions based on the material in your answer but your conclusions may be limited or unbalanced.

Level 3 (9–12 marks) You show relevant and accurate geographical knowledge and understanding of the role played by volcanic outgassing. You make sound connections between aspects of your answer as appropriate and support your interpretations logically with evidence. You draw balanced and logical conclusions based on the material in your answer.

Hints and tips

Remember to weigh up the various elements relevant here to come to an overall assessment at the end, based on the argument you have presented.

(e) 20 marks (AO1 = 5 marks, AO2 = 15 marks)

AO1 Demonstrating your knowledge and understanding

➤ There are three broad categories of alternatives to fossil fuels: renewables/recyclables, biofuels and radical technologies.

➤ Renewables are growing globally and meeting an ever-increasing proportion of global energy demands.

➤ Biofuels are also growing, especially in countries such as Brazil and the USA.

➤ As technology advances, new ways of capturing carbon are being found.

AO2 Applying your knowledge and understanding

➤ Despite the positive signs of the growth of global use of renewables, there are concerns about energy accessibility — not all countries have the same local access to the renewables.

➤ There are also social and environmental consequences of the use of renewables, including visual pollution, methane emissions from hydroelectric power (HEP) reservoirs, and social impacts such as the displacement of people.

➤ It is unlikely that most countries will be able to meet their energy needs through renewables alone, so some are turning to nuclear power as an option.

➤ Nuclear power has many benefits: low carbon emissions, limited air pollution, reliability; but there are also concerns — safety, security and disposal of waste.

➤ Biofuels are also lower emitters of carbon and the market for biofuel for transport is growing. However, there remain concerns, especially related to the displacement of other farming activities as the land needed for biofuels expands. If this land involves the clearing of forests, then are biofuels really carbon neutral?

➤ Alternative technologies include carbon capture. New technologies are being developed that make this increasingly effective and efficient. But concerns exist, including about the extra energy required at the power plants to actually capture the carbon — does this offset the benefits of capturing it in the first place?

➤ Hydrogen fuels also are an alternative form of technology. They are currently in the early stages of development, but they are non-polluting and there is optimism about the role they might play in meeting energy needs in the future.

Answers to this question will be given a mark within a level band

Level 1 (1–5 marks) You include isolated points of geographical knowledge and understanding of the merits of the alternatives to fossil fuels and their connection with global emissions, with some errors and inaccuracies. You have not made connections from the question to the points made. Your answer is incoherent and lacks relevant evidence to support ideas. Your argument is limited, with unbalanced points. Points that you make are concluded in a general manner, if at all.

Level 2 (6–10 marks) You make some points showing geographical knowledge and understanding of the merits of the alternatives to fossil fuels and their connection with global emissions, some of which may be relevant. You make some inaccurate points. You apply some of your knowledge but your ideas are not developed or may not be linked directly to the question. You use some evidence to support statements, which may answer only part of the question. You make a conclusion but this is drawn from often unbalanced ideas.

Level 3 (11–15 marks) You make generally relevant points showing geographical knowledge and understanding of the merits of the alternatives to fossil fuels and their connection with global emissions. Your ideas are mostly accurate and some connections are made between ideas. You interpret the question well in general but there may be some gaps in the use of evidence to support points. You draw a conclusion that links to the arguments made but is not fully supported by evidence.

Level 4 (16–20 marks) You show good use of geographical knowledge and understanding of the merits of the alternatives to fossil fuels and their connection with global emissions. You make a range of relevant points to create a coherent argument supported by appropriate evidence. You apply your knowledge well throughout. All points you make are linked to the question. You draw a good, well-balanced conclusion that links clearly to the evidence presented.

Hints and tips

Consider and weigh up the various alternatives, and come to an overall conclusion based on the argument you outline.

Question 2 example responses

Student A

(a) The graph shows that infiltration rates start higher but then quickly they start to fall off steeply, until they begin to level off to a lower and more constant rate. At the same time, the rate of surface runoff starts off lower and more constant, before starting to rise as the infiltration rate levels off.

These two flows are linked. The rate of infiltration starts to fall as the soil store is filling up with water. As it fills, the infiltration capacity of the soil drops. At the same time, surface runoff starts lower as most of the precipitation that falls is infiltrating. However, as the infiltration capacity falls over time, less water infiltrates and so it travels by surface runoff.

The student clearly identifies the trends in the graphs and successfully links the two flows in a clear and effective explanation. **3 marks**

(b) Storm hydrographs show how river discharge varies in response to a storm. As the rainstorm begins, the discharge rises slowly. There is little initial change during the rainstorm as most rain does not fall directly into the river. Soon, the rising limb of the discharge rises steeply towards its peak discharge. After reaching its peak, the falling limb falls and the discharge returns to its original level.

This hydrograph can be affected by various physical and human factors, making it either a flashy hydrograph (which has a short lag time and a high peak discharge) or a flat hydrograph (with a longer lag time and lower peak discharge).

First, the physical factors. In smaller basins, the precipitation has less distance to travel before it reaches the mouth, so the hydrograph will be shorter and steeper. Basin shape has an influence. Shorter, more rounded basins are more likely to be flashy as the water from the basin tends to arrive more quickly at the mouth. In a longer, thinner basin, the water that falls near the source has much further to travel and so it will produce a flatter hydrograph. Relief is important too. In steeper basins, under the influence of gravity, water will make its way to the mouth more quickly.

Soil type is another factor. Clay soils have much smaller pore spaces and so do not allow for much infiltration. As a result, overland flow is more likely and so the water reaches the channel quickly. The opposite is the case for sandy soils. When it comes to geology, some rocks, such as basalt, are impermeable and less infiltration occurs. This produces flashier hydrographs. More permeable rocks, such as chalk, allow infiltration and produce flatter hydrographs.

Second, human factors. Some land uses produce flashy responses. In urban areas, the impermeable surfaces increase runoff, and the drains and sewers are designed to take the surface water to the river quickly. Where vegetation is removed for agriculture, it leaves bare soil. This reduces interception and so the water gets to the channel more quickly. Other land uses produce flatter responses. For example, afforestation increases interception and thus slows down the speed at which the water reaches the channel. Furthermore, increased interception results in more evaporation, so the total amount of water reaching the channel is reduced, lowering the peak discharge.

(e) **A wide series of relevant factors are outlined and explained in detail.**   **Level 3, 6 marks**

(c) Flooding occurs when the discharge of a river overflows its banks. This is a perfectly natural process, and can occur as a result of various physical factors. One of these is rainfall. There are places in the world with very seasonal rainfall, such as the monsoon which affects the Indian subcontinent. For example, the River Indus flooded badly in 2010 following an intense monsoon. In July, the northwest region of Pakistan had 60 hours of continuous rainfall producing over 200 mm of rain. As a result, a huge flood peak of discharge started to travel south down the Indus into central Pakistan. River flow peaked at 32,000 m³/s. In addition to seasonal rain, there can be intense storms that lead to flash flooding. For example, the river flooding of the Mississippi in 2011 was caused in part by a series of four intense rainstorms in April in the Ohio River basin, which produced six times the average monthly rainfall for the area. This sent a flood peak down this tributary towards the confluence of the Ohio with the Mississippi at Cairo, Illinois.

A second physical cause of flooding is snowmelt. If snow falls onto high mountains, it can be stored there over the winter, only to be released again in the following spring as temperatures start to rise. Snowmelt contributed to the Mississippi flood of 2011. In the upper Mississippi and Missouri tributaries to the northwest of the Mississippi Basin, the snowfalls of winter 2010/2011 were record-breaking, with the snow being around 600 mm deeper than average. When snowmelt occurred in April, this meant that there was a higher quantity of water entering the river system. This raised the peak discharges and sent a flood peak down these tributaries that peaked at 15 m in Illinois.

Although it is a natural process, flooding can be made worse by human activities. One of these is changing land use. The Mississippi again illustrates this. There has been extensive deforestation in the Mississippi Basin since European settlers began to arrive in the eighteenth century. Fewer trees meant that there was less interception of rain when it fell, causing peak discharges to rise more quickly to higher levels. Furthermore, lack of trees meant that less of the water was taken up and outputted via transpiration, increasing the overall volume of water in the drainage basin. This increased the scale of the 2011 flood. A second human factor is urbanisation. The increase in urban areas increased surface runoff, resulting in faster and higher peak discharges, and reduced the storage capacity of the river basin, increasing the total volume of discharge in the river.

ⓔ **This extensive answer shows a detailed understanding of a wide range of relevant geographical factors throughout.** **Level 3, 8 marks**

(d) Volcanic outgassing is one of the processes whereby geological carbon is released into the atmosphere and is part of the long-term geological carbon cycle. Volcanic outgassing occurs at various tectonically active places, for example at tectonic plate margins, including subduction zones. One such volcano is Mount Pinatubo, which erupted in 1992, emitting 42 million tonnes of carbon dioxide into the atmosphere that had been brought to the surface in magma from its store in the asthenosphere below. The volcanic activity at constructive margins in mid-ocean ridges tends not to be as explosive and so does not emit as much carbon as in subduction zones. That said, given the extensive distribution of these margins, they still emit around 80 million tonnes of carbon dioxide per year, a significant proportion of the 200 million tonnes emitted annually by all volcanic activity.

However, the diagram shows that, over a long enough time period, volcanoes are part of the loop of the carbon cycle. This is because of a negative feedback loop set up by these carbon emissions. In the short term, the extra carbon in the atmosphere results in an enhancement of the greenhouse effect, resulting in rising temperatures — this extra atmospheric energy causes more global precipitation. The additional rain helps with chemical weathering, as the carbon dioxide mixes with the rainwater to form a weak carbonic acid. When this rain reacts with rocks, it dissolves them and produces calcium ions. These are then transported by rivers and deposited in layers elsewhere on ocean floors. The carbon is stored in these layers; in fact, over time, they can form into sedimentary rocks — calcite sediment forms into limestone.

Therefore, we can see that although volcanic outgassing causes short-term variation in climate through the release of carbon into the atmosphere, in the longer term, because of the negative feedback loop, it ultimately balances this emission out by contributing to the geological sequestering of carbon. However, over even longer time spans of millions of years, this rock can experience subduction and be carried into the asthenosphere, from where it can rise with magma into the lithosphere and be outgassed once again at volcanoes, starting the feedback loop over again once more.

However, it is important to realise that volcanic activity is not the only process affecting the carbon cycle on long timescales. Carbon is sequestered from the atmosphere by biological activity on land, and plant-based carbon can be stored as peat, and fossil carbon in the form of coal oil and gas. Volcanic emissions are just one part of the geological carbon cycle which encompasses carbonate sedimentary rock and geological carbon stored as fossil fuels.

ⓔ **The student shows good detailed and relevant geographical knowledge throughout. In addition, they use the figure well and combine that with their understanding to explain in detail the processes involved in the negative feedback loop. The conclusion is detailed and follows naturally and coherently from the earlier part of the answer.** **Level 3, 11 marks**

(e) There are three main categories into which you can put the alternatives to fossil fuels.

The first is renewable (wind, solar, geothermal, tidal and hydroelectric power (HEP)) and recyclable energy (nuclear). There is already widespread and increasing use of renewable energy. It is estimated that just over 10% of the world's global energy consumption comes from renewable sources. In 2015, nearly US$290 billion was invested in renewable technologies. There is some evidence of success in the uptake of renewable sources. For example, much of China's energy supply for its economic development has come from the burning of coal, which emits high levels of carbon dioxide. However, in recent years, China has been moving away from the use of coal. In 2015, its coal production and coal-fired energy production both fell by 3% and the government banned the development of new coal mines for 3 years. At the same time, low-carbon energy production sources increased by more than 20% in China. Stories like this have led some people to be optimistic about the future of renewables. In 2011, a projection by the International Energy

Agency suggested that solar energy could produce most of the world's electricity by 2060. Energy generation from renewables is certainly growing. In 2015, for the first time, renewable energy sources made up the majority of all new power capacity (54%).

However, others are less hopeful. Renewables like solar may well be growing and may well in the future supply considerable amounts of energy. That said, there remain issues with it — most notably related to physical availability of solar power: not all countries have enough hours of sunshine to locally generate sufficient power from solar energy. Indeed, not all countries have enough tectonic activity to generate geothermal energy. Not all countries have coastlines that will allow for tidal power. Consequently, it is unlikely that most countries will be able to rely on renewables alone to replace their current power needs. There are also environmental concerns with the use of some renewables. These include the visual pollution associated with wind farms (although some people do not mind these). Additionally, there are issues with the reservoirs needed for HEP dams. Reservoirs need to be large enough to store sufficient water to smooth out the annual variations in river regimes. This means that they can flood large areas, disrupting natural habitats found there. Some reservoirs in tropical areas can generate significant amounts of methane (a potent greenhouse gas) from rotting vegetation. One study has suggested that reservoirs can give off more greenhouse gases than an oil-fired power station if the trees have not been cleared before the valley was flooded. In addition, there are issues of efficiency (reservoirs can experience siltation as sediment is deposited there, reducing their efficiency) and the social impacts can be negative (the Three Gorges Dam in China displaced 1.24 million people).

Given the challenges that still exist with renewables, some countries look to recyclable energy in the form of nuclear power to meet their energy mix. On the positive side, nuclear power generates limited carbon outputs, between 50 to 80 times less than fossil fuel power plants. Globally, around 11% of electricity generated comes from nuclear power plants. The plants emit virtually no localised emissions, so there are no issues of air quality around them. They can generate recyclable and reliable energy to supply the baseload demands of a country (unlike some renewables). However, nuclear power is controversial and comes with various risks. There can be nuclear accidents, such as with the Fukushima plant in Japan which was damaged by the 2011 tsunami, causing nuclear contamination to be spread into the surrounding countryside and ocean. There are concerns about terrorism and plant security. There are issues with the disposal of the radioactive waste. Finally, there is the cost — nuclear power plants are expensive and so may not be a realistic option for the poorest countries.

This brings us to the second category: biofuels. In 2010, 3% of the world's transport fuel was from biofuels. In the USA, nearly all of the petrol sold is mixed with 10% ethanol, and car companies such as Ford and GM have developed flex-fuel vehicles that can run on petrol containing up to 85% ethanol. Brazil has also made extensive use of biofuels. Around 90% of all new passenger vehicles sold in Brazil have flex-fuel engines. The push for biofuel development here has meant that Brazil now is one of the leading exporters of sugar cane and ethanol, and the area of sugar cane cultivation was expected to double between 2003 and 2018. However, there are issues related to this expansion of cropland, as it has displaced other farm uses, especially cattle pasture. As a result, this has put pressure on rainforest land for cattle pasture instead. Deforestation not only contributes directly to carbon emissions (as the trees are often burnt to clear the land) but also removes part of the vital carbon store of the global carbon cycle. Thus, the environmental benefits of biofuels may be counteracted by these environmental losses.

The final category is the use of radical technologies, including carbon capture. The aim of carbon capture is to collect carbon dioxide from sources of emission (such as power stations) and transport it elsewhere where it can then be put into terrestrial stores or sinks so that it remains there rather than entering the atmosphere. The carbon dioxide is dissolved in water and injected into deep rock stores where it reacts with the rock to form carbonates.

One report suggests that there is capacity in North America for 900 years of carbon storage, based on current usage levels. In the UK, the depleted oil and natural gas fields in the North Sea could be used for carbon storage. In 2016 in Iceland, new technological approaches led to solid carbonates forming in 2 years (compared with the hundreds of years that had previously been predicted). This has led the Icelandic government to aim to bury 10,000 tonnes of carbon dioxide per year.

However, there remain concerns about the possibility that the carbon dioxide may leak back out into the atmosphere, perhaps in the more worrying form of methane. In addition, there are considerable extra energy costs involved in carrying out the capturing of carbon. For example, for coal-based power stations, the extra power needed ranges from 24% to 40%, significantly increasing the amount of fuel needed to produce the same amount of power.

In conclusion, it is clear that, although alternative energy sources are a vital part of transforming our global energy consumption away from an over-reliance on fossil fuels, each of the alternatives comes with its own issues and much careful management is needed to get the maximum benefit from them at the minimum economic, social and environmental cost.

ⓔ The student not only shows sound geographical knowledge throughout, with extensive and effective use of examples and figures, but also draws out relationships between positives and negatives, weighing them up against one another. A substantial and credible conclusion is reached. **Level 4, 20 marks**

Student B

(a) The infiltration rate on the graph starts high but then quickly falls off, dropping steeply. It then levels out to a lower and more constant rate. Meanwhile, the rate of surface runoff starts low and more constant and then rises as the infiltration rate levels out.

ⓔ The student clearly identifies the trends in the graphs but does not link the two flows to provide an explanation. Read the question carefully and answer precisely what is asked.

2 marks

(b) A storm hydrograph shows how the river discharge varies in response to a storm. A hydrograph can be either flat (with a short lag time and a high peak discharge) or flashy (with longer lag times and lower peak discharges) depending on various physical and human factors.

ⓔ The student mixes these two up. Mistakes like this can happen in the exam when under pressure, but take care not to make careless mistakes so that your answer can earn maximum marks.

Relief is an important physical factor. For example in steeper basins, the influence of gravity means the water makes its way to the mouth of the river more quickly. The hydrograph will be shorter and steeper in smaller basins, as the precipitation has less distance to travel before it reaches the mouth. Soil type is another physical factor. Overland flow is more likely with clay soils as they have much smaller pore spaces which don't allow for much infiltration, and so the water reaches the channel quickly.

Human factors also affect storm hydrographs. A flashy response is seen with some land uses, for example urban areas. Impermeable surfaces here increase runoff and the drains and sewers take the surface water to the river quickly. On agricultural land, bare soil is left when vegetation is removed so interception is reduced and water gets to the channel more quickly.

A flatter response is caused by other land uses such as afforestation. Interception is increased here, slowing down the speed at which the water reaches the channel. The peak discharge is lower as increased interception also results in more evaporation, so the total amount of water reaching the channel is reduced.

(e) **Overall, there are some relevant factors. But the student makes one error in understanding and the factors need to be developed more for a top-level response.** **Level 2, 3 marks**

(c) Flooding can occur due to different physical factors, such as rainfall. There are places in the world with very seasonal rainfall, such as the monsoon which affects the Indian subcontinent. The monsoon affects this region in the summer months and can drop significant amounts of rainfall in a few short months, overwhelming the river banks. You can also have flooding that follows intense rainstorms in the UK. For instance, in 2015, an intense rainstorm over Cumbria caused significant flooding there, flooding nearly 5,000 homes. The intense rain means that the infiltration capacity is exceeded and more surface runoff occurs.

Flooding can also be caused by snowmelt. For example, the Indus River in Pakistan is fed by seasonal melting of the glaciers and snowpack. This is one of the reasons for the devastating flood there in 2010.

Human factors also cause flooding. For example, land use change. Again in the Indus River, there has been extensive deforestation — the country's forest cover will be reduced to half of its 1995 level by 2020. With the trees removed, there is less interception of the rains, meaning that both the volume and the speed of water reaching the rivers increase, contributing to the exceptionally high flood peak levels. Hard engineering can also contribute to flooding. The levées built up beside the Indus to reduce the flood risk actually made it worse. By cutting the river off from the floodplain, the levées meant that sediment was kept in the channel and deposited on the river bed, raising its level. The levées then failed during the flood and the waters spilled out onto the floodplain.

(e) **The student refers to a range of physical and human factors and makes good use of examples to illustrate their answer. It could be improved by showing a fuller understanding of exactly how these factors link to flooding. For instance, in the first paragraph, these connections are not made as clearly as they might be.** **Level 3, 6 marks**

(d) Volcanic outgassing happens at various tectonically active places, such as tectonic subduction zones. For example, Mount Pinatubo, which erupted in 1992, emitted significant amounts of carbon dioxide that had been brought to the surface in magma, into the atmosphere. The volcanic activity at constructive margins in mid-ocean ridges tends not to be as explosive and so does not emit as much carbon as in subduction zones. Outgassing also occurs at hot spots such as Hawaii and at places like Mount Etna, a collision zone on Sicily. This clearly adds carbon into the atmosphere.

However, this carbon can set in motion a negative feedback loop which ultimately acts to return carbon to the geological store, when measured over the timescale of hundreds of thousands of years. It happens like this. The extra carbon in the atmosphere initially leads to an enhancement of the greenhouse effect. In the short term, this first of all leads to an increase in precipitation. But this precipitation mixes with the carbon dioxide to produce weak carbonic acids. These react with the rocks on the surface, causing chemical weathering, releasing ions from the rocks. These are then carried away by the rivers and deposited on ocean floors. This way, the carbon is taken back into the geological store.

So we can see how, though outgassing releases carbon in the short term, over longer timescales it can bring carbon back into the geological carbon store via this negative feedback loop.

(e) **The student shows good knowledge and understanding, but could be more detailed in places to earn more marks. For example, more details on the places and examples referred to could be given. In addition, the conclusion is reasonably coherent, but again more detail would show a deeper understanding of the longer-term outgassing cycle.** **Level 2, 7 marks**

(e) The first alternative to fossil fuels is renewable and recyclable energy. There is already widespread and increasing use of renewable energy. For example, much of China's energy supply for its economic development has come from the burning of coal, which emits high levels of carbon dioxide. However, in recent years, China has been moving away from the use of coal. In 2015, its coal production fell and low-carbon energy production sources increased by more than 20%. Stories like this have made some more positive about the future of renewables. An estimate by the International Energy Agency suggested that solar energy could produce most of the world's electricity by 2060.

Not everyone is as positive, though. For example, there are issues with solar power. Not all countries have enough hours of sunshine to locally generate sufficient power from solar energy. Indeed, not all countries have coastlines that will allow for tidal power. Consequently, it is unlikely that most countries will be able to rely on renewables alone to replace their current power needs. There are also environmental concerns with the use of some renewables. These include the visual pollution associated with wind farms (although some people do not mind these). This is known as NIMBYism (Not In My Back Yard). If there are proposals to locate a wind farm nearby, then people may often protest. There were plans to put an offshore wind farm near the AONB at the Giant's Causeway, Northern Ireland, but these were dropped after opposition from environmentalists.

Some countries are also using recyclable energy including nuclear power. This power source emits little carbon and it is reliable and can help meet the baseload needs of a country. But there are concerns about nuclear power. There can be nuclear accidents, such as with the Fukushima plant in Japan which was damaged by the 2011 tsunami. There are concerns about terrorism and plant security. There are issues with the disposal of the radioactive waste. Finally, there is cost — nuclear power plants are expensive and so may not be a realistic option for the poorest countries.

Biofuel is another alternative to fossil fuels. Brazil has made extensive use of biofuels and is now one of the leading exporters of sugar cane and ethanol. The area of sugar cane cultivation in the country was expected to double between 2003 and 2018. However, there are issues related to this expansion of cropland, as it has displaced other farm uses, especially cattle pasture. This has put pressure on rainforest land for cattle pasture instead. Deforestation removes part of the vital carbon store of the global carbon cycle.

Radical technologies can also be used. One of these is carbon capture. The aim of carbon capture is to collect carbon dioxide from sources of emission and transport it elsewhere where it can then be put into terrestrial stores or sinks so that it remains there rather than entering the atmosphere. It is thought that there is capacity in North America for 900 years of carbon storage. In the UK, the depleted oil and natural gas fields in the North Sea could be used for carbon storage. On the other hand, carbon capture is expensive and uses more energy in the process of capturing it. Hydrogen fuel cells are another alternative technology. These cells are non-polluting as they only emit electricity, heat and water. They can be used to power cars and this technology seems to have a good future ahead of it.

Overall, there are some positives and some negatives in the use of alternatives to fossil fuels.

e **This student covers a range of alternatives with relevant material. More marks could be gained by making more connections between parts of the answer. For example, the reason why some countries are exploring the use of recyclable energy is due to issues with the effectiveness of renewables — this link could be made more explicit in the paragraph about nuclear power. Also, the conclusion is too brief and not really informed by the discussion that came before it.**

Level 3, 12 marks

Area 4 Human systems and geopolitics

Topic 7 Superpowers

Question 1 mark scheme

(a) 4 marks (AO1 = 4 marks)

This question asks you to demonstrate your knowledge and understanding of the role that intergovernmental organisations (IGOs) have in encouraging and maintaining geopolitical stability between different countries of the world. You gain 1 mark for giving a reason for the global geopolitical importance of IGOs. The remaining 3 marks can be gained by developing examples of the different areas of interest of IGOs and how these contribute to global stability (for example through promoting peace and security, resolving conflicts and reducing the impacts of climate change). To gain maximum marks, you should show good knowledge and understanding of the different roles that an IGO has in the global geopolitical system, supported by named examples, e.g. UN Security Council, International Court of Justice.

For example:

IGOs are organisations in which two or more countries work together on areas of interest to all parties. They play a vital role in promoting global peace and security because they can create a system where the world's people can work together on international issues (1) and where problems can be discussed and solutions to differences put forward (1). An example of an IGO is the United Nations. The United Nations promotes international cooperation (1) through many different bodies. The UN Security Council is responsible for putting forward resolutions to promote global peace and security, as well as sending UN peacekeepers to areas of ongoing conflict (1). The International Court of Justice is responsible for settling legal disputes between member states such as those linked to military or resource disputes. However, not all nations are members of IGOs and so can be excluded from the decision-making process.

Other appropriate reasons will be credited.

Hints and tips

Think about an example of an IGO and the areas that it can help promote stability (e.g. political/environmental challenges).

....

(b) 12 marks (AO1 = 3 marks, AO2 = 9 marks)

This question focuses mainly on applying your knowledge and understanding of how trying to obtain physical resources, such as oil and gas, can lead to disputes over ownership and exploitation. Superpowers have different attitudes towards obtaining natural resources and their actions may lead to disagreements and conflicts. Relevant points that you could mention and expand are suggested below.

AO1 Demonstrating your knowledge and understanding

➤ Superpowers have exploited physical resources in areas that are in disputed territory.

➤ Advances in technology and climatic change have increased opportunities for physical resources to be exploited.

- Disagreements exist between nations as to who has the right to exploit resources.
- The exploitation of resources by superpowers may cause tension with other groups, e.g. NGOs.

AO2 Applying your knowledge and understanding

- Ownership disputes may be caused by historical and contemporary territory disagreements.
- New areas may become available for exploitation because of landscape and/or seascape changes resulting from climatic change or technology availability.
- Increased demand for physical resources or need to reduce energy dependency can lead to exploration of new locations for exploitation.
- Governments may come into conflict with other interest groups, e.g. NGOs, when they exploit areas for physical resources.

Answers to this question will be given a mark within a level band

Level 1 (1–4 marks) You show only a limited knowledge and understanding of the tensions that can arise when physical resources are obtained by superpowers. The examples you use to support statements may be very general and lack case study detail. At the bottom end of this band you show no attempt to assess the reasons for disputes and disagreements.

Level 2 (5–8 marks) You generally show a good understanding of the tensions that can arise when physical resources are exploited by superpowers. You show understanding of ownership disputes and disagreements over exploitation, with some examples used to support statements. You make some attempt to assess reasons for conflicts of interests. You show some consideration of the relationship between superpowers, as well as the relationship between superpowers and other organisations.

Level 3 (9–12 marks) You show accurate knowledge and understanding of the tensions that can arise through exploitation of physical resources by superpowers. You apply your knowledge to effectively assess a range of causes of tensions and to make a judgement of their significance. At the upper end of this band, you confidently use a range of detailed case study information in your supporting statements.

Hints and tips

Can you give examples of how resource exploitation in a particular area has led to tensions? Why have these tensions arisen? Who is involved?

Question 1 example responses

Student A

(a) IGOs are made up of groups of countries which share common interests. IGOs are important to geopolitical stability because they allow the governments of different countries to <u>discuss issues</u> which affect them and to try and <u>resolve disputes and conflicts</u> between them. Such <u>problems include</u> disagreements about territories and borders, the use of resources, asylum issues and maritime disputes. The UN is the world's major IGO with over 190 member states and has institutions within it which deal with specific areas of international interest. For example, the <u>UN Security Council</u> can provide a neutral presence through its peacekeeping missions in areas of longstanding conflict. This helps to resolve problems which have been difficult to deal with in the past. The UN can also act to challenge countries that have caused disputes with other member states through setting sanctions or authorising military action. IGOs can also tackle environmental issues such as climate change, which affect the international community.

ⓔ This is a good answer, giving clear reasons for the importance of IGOs in relation to geopolitical stability. Different areas of international interest are discussed in general, and the UN example is used to support points.

4 marks

'discuss issues' Reason identified.

'resolve disputes and conflicts' Second reason identified.

'problems include' Expansion of relevant issues.

'UN Security Council' Relevant example to support points.

(b) Demand for physical resources such as gas and oil has increased because of population and economic growth, as well as the need to improve energy security. The search for new locations for resource exploitation has continued, but developing these areas can lead to tensions and conflicts between superpowers themselves, as well as with other groups. The Arctic Ocean is one of the last wilderness areas and is under threat from oil and gas exploration, as it is estimated that 30% of the world's new gas reserves and 13% of its undiscovered oil are found within the Arctic Circle. The Arctic region itself is bordered by two superpowers, Russia and the USA, as well as Canada, Denmark, Norway and Iceland, and includes international waters. Resource exploitation may result in disagreements between these nations as they apply for the right to drill for oil and gas by making claims for territory under a UN Law of the Sea Treaty. The number of locations where drilling could be possible in the Arctic has increased due to the effects of global warming. The number of ice-free days in the Arctic Ocean is increasing, allowing further resource exploration to take place. This may lead to increased conflicts with NGOs, such as Greenpeace, that are concerned with biodiversity and nature conservation, as the Arctic ecosystem may be negatively affected by the drilling. Superpowers like Russia may start to increase their military presence in the Arctic, which could lead to other countries doing the same. TNCs such as Shell Oil may become involved with the development of Arctic oil reserves as they are able to invest in the processes needed to exploit this technically difficult location. TNCs need to work with governments for drilling to be allowed but this could lead to tensions with local communities. Overall, the need to obtain physical resources is a significant cause of disputes between superpowers such as Russia and the USA. As demand for resources in emerging superpowers such as China increases, conflicts may become more complex. Therefore, there may be an increasing need for IGOs, such as the UN, to settle disputes over territory and take measures to avoid future conflicts.

ⓔ The student shows good knowledge of the reasons for tensions caused by the exploitation of physical resources and applies their knowledge well to the Arctic Ocean case study. A range of players involved with disputes and disagreements is suggested, with the focus on superpowers. The Arctic case study is well used to support the main points and a judgement is made in the final paragraph to suggest that other nations and IGOs may become increasingly involved. A discussion of tensions in another region, e.g. gas and oil reserves in the South China Sea, would have further supported the answer and potentially increased the marks.

Level 3, 10 marks

Student B

(a) The world has many problems and sometimes it is difficult to sort these out. IGOs help different countries to come together to talk about the problems that they may have between them. IGOs help countries to work out how these issues <u>can be solved</u> and this can help the world to be a more peaceful place. Two examples of IGOs are the <u>UN and the G8</u>.

(e) **The student shows some understanding of the role of IGOs but does not give details of how they achieve their goals. Reference is made to examples, but marks are limited by lack of detailed explanation of how IGOs try to achieve geopolitical security.** **2 marks**

'can be solved' Role of IGOs stated, with general reference to security.

'UN and the G8' Examples given, but no discussion of contribution to geopolitical stability.

(b) The USA and Russia are both trying to drill for oil and gas in the Arctic Ocean. These superpowers need more resources to help develop their economies. This means that they compete with each other to find new places which have oil. They can come into conflict with each other when they do not agree about who owns the area where oil is found. International organisations like the UN have to intervene to help nations decide who owns what and some governments might not agree with their decisions. Some charities like the WWF may be against exploitation of oil because the equipment used will harm the fragile ecosystem in the Arctic. They may feel that their views will not be heard and better technology may mean more oil is extracted from areas already under threat. Another area where different countries are trying to find and drill for resources is the South China Sea. China is a developing superpower and needs more oil and gas as its economy is growing and it is expensive to import energy from other countries. China needs to find new places where it can get oil. It has already begun drilling in the South China Sea and other countries disagree with this. If IGOs do not help to solve these disputes, the area may be affected by conflicts in the future.

(e) **Some reasons for tensions surrounding resource exploitation are identified but could have been developed in more detail. Supporting statements are linked to two specific examples, the Arctic region and the South China Sea, but the issues are discussed only in a broad context; marks could have been gained by discussing the geopolitical circumstances of resource exploitation. The answer focuses on the role of superpowers, but other relevant groups that may have an interest in highlighting environmental and community issues or an involvement in resolving disputes are mentioned. A judgement on how significant resource availability is as a cause of disputes among superpowers is needed.** **Level 2, 7 marks**

Question 2 mark scheme

(a) 4 marks (AO1 = 4 marks)

This question focuses on showing your knowledge and understanding of the reasons why emerging countries such as Brazil, Russia, India and China (BRIC) and other G20 members play an increasingly important role in global economic and political affairs. You should explain factors such as economic growth, population size and structure, land area, physical resources, role in intergovernmental organisations and global political influence, where appropriate, for different emerging countries. You gain 1 mark (up to a maximum of 3) for each reason that explains the importance of emerging countries' roles in the global economic system, and 1 mark (up to a maximum of 3) for each reason that explains the importance of emerging countries' roles in the global political system. A maximum of 4 marks is available for this question.

For example:

BRIC countries are becoming increasingly important on both the global economic and political stages. This is because over 40% of the global population lives in BRIC countries (over 2.5 billion people in India and China combined) (1) and their combined GDP is around 20% of the global total (1). This means that, with increasing standards of living and wealth, a rising middle-class population is becoming an important market for both domestic and international businesses (1). BRIC countries also have a large amount of physical resources such as oil and gas (Russia) (1) and can produce high qualities of manufactured goods (e.g. 40% of China's exports are mechanical and electrical products) (1). The export of these primary and secondary goods by the BRIC countries can have a significant effect on the global trading system (1). The BRIC nations also influence global financial policy through their G20 and IMF memberships (1). BRIC countries play an important role in global decision-making. They are longstanding UN members, with China and Russia influencing global political affairs by being members of the UN Security Council (1). Some BRIC countries are also looking to increase their global political influence through developing their power in regions such as the Arctic (Russia) and the South China Sea (China) (1).

Other appropriate reasons will be credited.

(b) 12 marks (AO1 = 3 marks, AO2 = 9 marks)

AO1 Demonstrating your knowledge and understanding

➤ Westernisation is a process where western (e.g. European, US, Canadian) ways of working, customs, lifestyles and technologies are practised or copied in other parts of the world such as emerging economies or less developed countries.

➤ Westernisation can have an influence on the global economic system.

➤ Other cultural processes may also be considered to be powerful or growing in power, such as Chinese (Sinofication) and Muslim (Islamification).

AO2 Applying your knowledge and understanding

➤ The spread of western customs, lifestyles, technology and working styles has influenced the global economic system with benefits and costs to a range of global players (e.g. societies, governments, businesses).

➤ Westernisation can also have an impact on both the economies of the 'western' countries themselves and the wider world.

➤ While westernisation is a powerful force within the global economy, other cultural processes may have a similar or greater influence in some parts of the world and on some groups of people.

Answers to this question will be given a mark within a level band

Level 1 (1–4 marks) You show only a limited knowledge and understanding of how westernisation can influence the global economic system. The examples used to support your statements may be very general and lack case study detail. At the bottom end of this band, you show no attempt to assess the extent to which westernisation influences the global economic system.

Level 2 (5–8 marks) You generally show a good understanding of how westernisation has an influence on the global economic system. You make some attempt to assess the extent of the influence of westernisation and start to put forward some ideas about how westernisation can affect the global economy, including the range of players within it. The examples you use to support statements are relevant. You start to consider how other processes or cultures may influence the global economic system.

Level 3 (9–12 marks) You show accurate knowledge and understanding of how westernisation has an influence on the global economic system. You consider its effects on a range of players within the system and put forward areas where westernisation has different levels of influence. You also discuss how other regions with different cultural processes may have an influence on the global trading system, e.g. China, the Arab world. You make a judgement as to the level of importance of westernisation. At the upper end of this band, you confidently use detailed case study information in your supporting statements.

Hints and tips

Think about what westernisation is and how it influences the global economy. How does the 'western' way of working, technology and culture affect businesses? Are there other cultural groups that are increasing in power in the global economic system in our world today?

Question 2 example responses

Student A

(a) Emerging countries are ones which are increasing their global economic power through rising levels of GDP, higher levels of economic growth than some western economies (e.g. China has had approximately 8% growth over the last 5 years) and large populations (1 in 7 people in the world lives in China). High levels of resource exploitation and manufacturing enable these countries to export their goods within the global trading system and to earn money to develop their own economy through infrastructure and social projects such as schools. They also have an increasing influence over global financial issues through their membership of the G20. Individually, they have the capacity to negotiate bi- and multinational trade deals to gain resources that they need (e.g. China offering money to fund Angolan infrastructure projects in return for oil). Politically, BRIC countries have very large land masses and this can help them to exert political power. For example, Russia's long northern coastline leaves it strategically well placed to influence the Arctic region. As members of the UN Security Council, Russia and China can play a significant part in the decisions relating to resolving conflicts around the world.

ⓔ **This is a good answer. A wide range of clearly explained reasons why emerging countries have increasing economic and political influence is given. Relevant examples from BRIC countries are used to help justify points. The student demonstrates a good range of knowledge and includes at least two economic and two political reasons.** **4 marks** 'global economic power' Clear links to economic systems.

'export their goods' Links to economic influence.

'membership of the G20' Links to global economic decision-making.

'China offering money' Example of influence on trading systems.

'strategically well placed to influence the Arctic region' Example of political influence.

'members of the UN Security Council' Example of influence on global decision-making.

(b) Westernisation is the way in which different customs and practices, such as those linked to lifestyle, culture, business methods and technology from the USA and Europe, have had an influence on other cultures. In many respects, it has had a hugely important influence on the global economic system. Much international business is conducted in English and this has increased with the ease of working using the internet. Information can be spread more quickly as the world can be connected through the internet, a 'western' invention, 24 hours a day. Many TNCs such as Coca-Cola and General Motors are American-based companies that spread their style of working practice and organisation into other countries through their branch plants in developing economies. Western-style clothing such as jeans now appears across generations and in many different cultures. However, while the style is western, jean production often takes place in emerging economies rather than those in the West. Thus 'westernisation' can have a negative impact on levels of textile manufacturing in the West. While western culture and ideas can seem to dominate many parts of the world, other cultures can be seen to be spreading their ideas. The increasing use of social media in the Islamic world has partly driven the growth of movements such as the Arab Spring and can be linked to radicalisation in Europe. These political changes can affect the stability of the global economic system. The rise of Chinese investment and workers in some Asian and African economies has changed working practices in some industries, e.g. road construction in Ethiopia and the gambling industry in Laos. Overall, westernisation has a significant influence on the global economic system. However, other cultural processes from different regions are having an increasing influence on parts of the global economic system.

ⓔ **This is a good answer. The student has shown how western ideas and practices strongly influence technology and business, and uses relevant examples to justify points. They also suggest how westernisation has an impact on the western countries themselves, in this case a negative one. How other cultures are spreading their ideas and businesses globally and how this may affect parts of the global economy are also discussed. A judgement is made on the level of influence of westernisation. To gain more marks, the student could discuss the influence of westernisation on other industries such as the sports business.**

Level 3, 10 marks

Student B

(a) Some large countries have increasing importance within the world. They can produce things that other countries want to buy and so are an important part of the trade system. For example, India produces a lot of <u>textiles which are then sold abroad</u>. This makes money for the Indian economy. India has a very <u>large</u> population. While some of these people are very poor, there are more and more people who are getting richer and are <u>able to buy more</u> products like smartphones. This helps companies like Apple increase their profits.

ℯ **The student shows some understanding of the question and uses examples from India to help explain how an emerging country can influence the global economic system. However, the answer does not include any reasons why emerging countries are important to the global political system so no marks are gained for this part of the question.** **2 marks**

'textiles which are then sold abroad' Example of emerging economy and link to global trade.

'large' Not sufficiently specific — population size? Land area?

'able to buy more' Increasing market within emerging economies, from which TNCs can benefit.

(b) The global economic system is influenced by many different organisations and governments. Traditionally, western countries such as the USA and the UK have dominated global trade because they have strong links with the rest of the world as many other countries want to buy their products. In the past, this was mainly manufactured goods but now other countries such as China make a lot of products to export. Western countries are significant in the financial markets. This is because they are well established and deal with the richest companies in the world. However, more Chinese companies, such as Huawei, which produces smartphones, are becoming more wealthy and are starting to have an impact on the world market. Western products are highly sought-after by more wealthy people in emerging economies. Companies like IKEA, Starbucks and Louis Vuitton have branches in countries outside of the West but sell many of the same products that they would at home. Some western companies adapt their products to increase their profit in other countries. For example, McDonald's offers a rice bowl and chicken in its Indonesian stores.

ℯ **The student shows some understanding, stating that western countries have dominated the global trading system and suggesting that other countries are now becoming more influential. This point could be expanded to suggest how China is also impacting on the global economic system by investing in other countries and importing primary products, which would have gained more marks. Examples of how western companies can influence the culture of a country by promoting products relating to design, fashion and food are given. A brief discussion of the influence of the internet would gain further marks.** **Level 2, 7 marks**

Topic 8 Global development and connections

Option 8A Health, human rights and intervention

Question 3 mark scheme

(a) (i)   3 marks (AO3 = 3 marks)

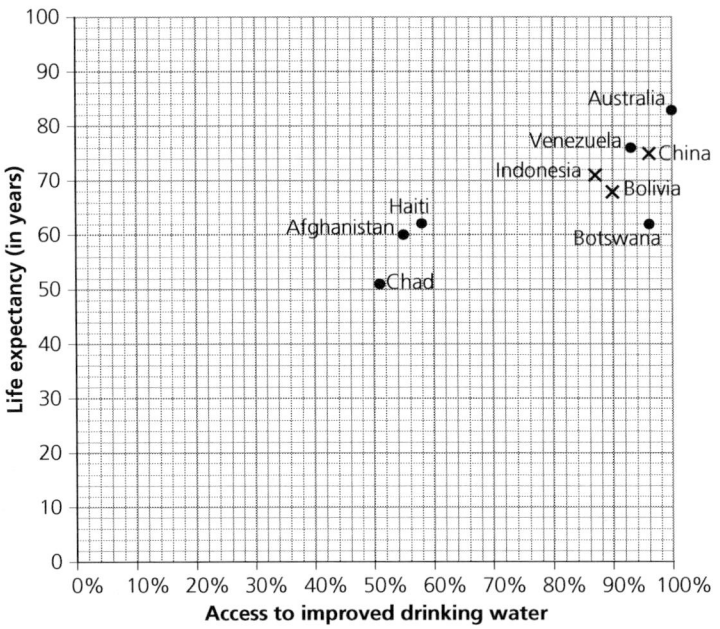

You gain 1 mark for each correctly plotted point for Bolivia, China and Indonesia.

(ii) 1 mark (AO3 = 1 mark)

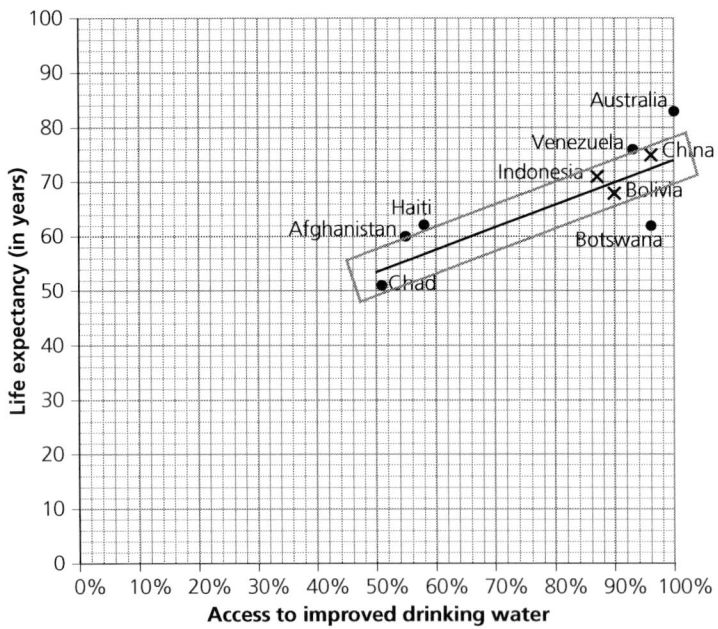

The best fit line for the graph is drawn in the centre of the rectangle. Parallel lines within the rectangle will be credited with 1 mark.

(b) 6 marks (AO1 = 3 marks, AO2 = 3 marks)

This question examines the relationship between life expectancy and percentage access to improved drinking water. You should be able to recognise that there is a weak positive correlation between the variables, i.e. as percentage access to improved water increases, so does life expectancy. Relevant content that could be included is suggested below. You do not need to include all of it in your answer and other relevant material can be credited.

AO1 Demonstrating your knowledge and understanding

➤ Percentage access to improved water shows the availability of clean water that is less likely to carry diseases which would affect life expectancy.

➤ More developed countries are likely to have better access to improved drinking water. Good quality water is likely to be more available to most of the population through established infrastructure and regulated companies. This will help to increase life expectancy.

➤ Less developed countries are likely to have poorer access to improved drinking water. Infrastructure may be poor or non-existent in some areas. This can lead to reduced life expectancy.

AO2 Applying your knowledge and understanding

➤ The relationship shown, with the limited data selection, is a weak positive correlation between the two variables. The selected countries can be divided into three groups: those with low-level access to improved drinking water and low life expectancy, e.g. Chad; those with higher levels of improved drinking water and medium life expectancy, e.g. Bolivia; those with both high access to improved drinking water and high life expectancy, e.g. Australia.

➤ Access to improved drinking water may be affected by water scarcity, lack of investment in infrastructure, overcrowding, pollution, climatic conditions, poor sanitation, overuse of wells, lack of adequate water treatment.

➤ Life expectancy can be affected by access to improved drinking water although it is linked to other variables, e.g. food availability and nutritional level, access to medical care, infant mortality rates, lifestyle.

➤ Countries with low-level access to improved drinking water and low life expectancy may have limited financial resources to invest in water supply improvements.

Answers to this question will be given a mark within a level band

Level 1 (1–2 marks) You show some general geographical knowledge and understanding of the relationship between life expectancy and percentage access to improved drinking water but some points are inaccurate. Your knowledge has not been applied consistently with the question. Some links that you have made between the graph and the question may be irrelevant.

Level 2 (3–4 marks) You show mostly relevant general geographical knowledge and understanding of the relationship between life expectancy and percentage access to improved drinking water. You make some relevant links between the graph and the question.

Level 3 (5–6 marks) You show accurate and relevant geographical knowledge and understanding of the relationship between life expectancy and percentage access to improved drinking water. You make logical connections between the graph and question.

Hints and tips

Can you explain the relationship? Can you group the countries? Think of social, economic and environmental factors that may affect the relationship.

(c)   8 marks (AO1 = 8 marks)

This question examines the way in which international agreements can be used to tackle human rights issues. These agreements are drawn up by a range of different organisations and can provide a framework to protect and promote human rights. Suggested ideas are outlined below, but you do not need to include all of these in your response. Other relevant points will be given credit.

AO1 Demonstrating your knowledge and understanding

➤ Human rights form the basis for freedom, peace and justice and are included in many different international agreements.

➤ International agreements can provide a universally standard definition for human rights which can be used in disputes between countries.

➤ The United Nations Universal Declaration of Human Rights was signed in 1948 and was designed to give a framework so that everyone can understand what human rights are.

➤ The European Convention on Human Rights focuses on human rights and freedoms within Europe. It also aims to promote gender equality and civil liberty and it established the European Court of Human Rights.

➤ International agreements on human rights can have an effect on development and international aid may be dependent on the human rights record of a country.

➤ International agreements on human rights can help to improve international peace and global security, e.g. refugees' rights during conflicts.

Answers to this question will be given a mark within a level band

Level 1 (1–2 marks) You show limited geographical knowledge and understanding of the role of international agreements in promoting human rights with some inaccuracies and a lack of geographical ideas.

Level 2 (3–5 marks) You show some relevant geographical knowledge and understanding of the role of international agreements in promoting human rights. You show a range of geographical ideas but your answer lacks detail.

Level 3 (6–8 marks) You show good geographical knowledge and understanding of the role of international agreements in promoting human rights. Your points are accurate, relevant and include a range of geographical ideas which are put forward in detail.

Hints and tips

Can you give examples of international agreements concerning human rights? Why are such agreements needed? Is their role positive and/or negative? Why?

(d)   20 marks (AO1 = 5 marks, AO2 = 15 marks)

This question focuses on the idea that aid does not always benefit developing countries. While aid can contribute towards projects that can improve aspects of the economy, society and the environment, it can have negative effects on developing countries. Aid can come from a range of different sources and in many different forms. Suggested ideas are outlined below, but you do not need to include all of these in your response. Other relevant points will be given credit.

AO1 Demonstrate your knowledge and understanding

➤ Development aid is assistance given to developing countries in order to meet some of the challenges they face.

➤ Development aid can be given in the short, medium or long term by a range of different groups, e.g. international organisations such as the IMF or NGOs such as Save the Children.

➤ Development aid can come in many different forms, such as money, expertise, equipment, training and technology.

➤ The success of development can be measured in a number of different ways.

AO2 Applying your knowledge and understanding

➤ The impact of development aid is disputed. Its effects on development can be positive for developing countries but can also have negative impacts.

➤ Bilateral aid is given from one government to another, often for large-scale projects such as hydroelectric dams. However, it is often seen as 'tied aid', where the developing country may have to buy products from the donor country in return.

➤ Multilateral aid is given by a group of countries or an international organisation such as the UN, IMF or World Bank. Loans and highly indebted poor countries (HIPC) initiatives can lead to increased debt and other economic problems for the developing country and contribute to uneven development within a country.

➤ Emergency aid is provided by governments and NGOs to people who have suffered natural disasters, e.g. earthquakes/hurricanes, or humanitarian disasters, e.g. civil war.

➤ NGO/charity aid is money given by the voluntary sector and may focus on a specific issue within a country. While money can be targeted to reduce a particular problem, funds may not be adequate or reliable enough to address the issue and projects may only help small groups.

Answers to this question will be given a mark within a level band

Level 1 (1–5 marks) You make isolated points of geographical knowledge and understanding of how development aid can lead to problems and solutions for developing countries with some errors and inaccuracies. You show limited understanding and you are not able to make connections. Your answer is incoherent and lacks relevant evidence to support ideas. Your argument is limited, with unbalanced points. Your ideas are concluded in a general manner, if at all.

Level 2 (6–10 marks) You show geographical knowledge and understanding of how development aid can lead to problems and solutions for developing countries, some of which may be relevant. You make some inaccurate points. You apply some knowledge, but your ideas are not developed or may not be linked to the question. You use some evidence to support statements which may answer only part of the question. You make a conclusion but this is drawn from often unbalanced ideas.

Level 3 (11–15 marks) You show geographical knowledge and understanding of how development aid can lead to problems and solutions for developing countries. Your ideas are mostly relevant to the question and you make accurate points. You make some connections between ideas. You interpret the question well in general but there may be some gaps in the use of evidence to support your points. You draw a conclusion which links to the arguments you make but may not be fully supported by evidence.

Level 4 (16–20 marks) You show good use of geographical knowledge and understanding of how development aid can lead to problems and solutions for developing countries. You make a range of relevant points to create a coherent argument supported by relevant evidence. All of your points are linked to the question. You make a good, well-balanced conclusion which links clearly to the evidence presented.

Hints and tips

Don't forget to evaluate the statement. You should give examples of successful and not so successful development aid projects. Present a balanced argument and make a judgement in your conclusion.

Question 3 example responses

Student A

(a) (i)

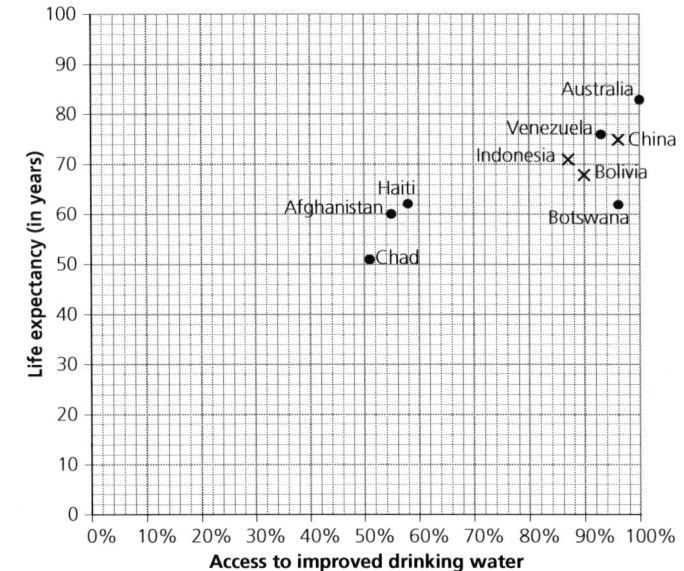

ⓔ The student has accurately plotted the data for Bolivia, China and Indonesia.   **3 marks**

(ii)

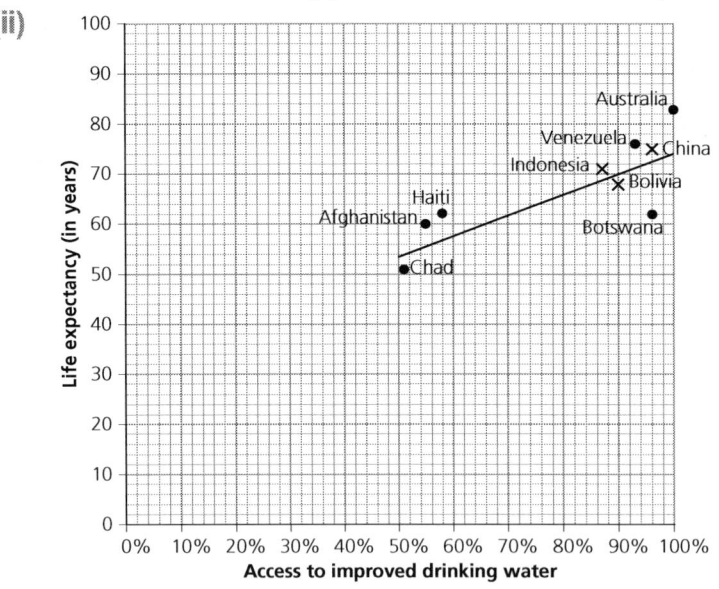

ⓔ The student has accurately drawn the best fit line.   **1 mark**

(b) Although the amount of data used is limited, we can see a slightly positive relationship between percentage access to improved drinking water and life expectancy. This shows that, in general, as access to good drinking water increases, so does life expectancy. Poorer countries such as Afghanistan and Chad have been affected by conflict which will have influenced the amount of water infrastructure. There may be large numbers of people in rural areas and it is more expensive to ensure that these communities have access to good drinking water. Also, high infant mortality rates have a negative impact on life expectancy figures in these areas. Some countries have greater access to good drinking water but a medium life expectancy. This could be because of the difference in infrastructure in urban and rural areas. For example, in China millions of people live in large cities which have recently developed their water systems, increasing water quality and improving health and increasing life expectancy. However, in some rural regions the development has been slower and communities may have to rely on water which is not as clean, negatively affecting health. Developed countries like Australia have high life expectancy and this is due partly to their well developed infrastructure for drinking water.

(e) **This is a good answer that identifies the key relationship in the graph. The student attempts to group countries together to explain differences both within and between groups. Connections between the graph and the student's knowledge are clear.** **Level 3, 6 marks**

'slightly positive relationship' Relationship identified.

'affected by conflict' Reason suggested and linked to graph.

'high infant mortality rates' Reason expanded to show understanding.

'greater access to good drinking water' Reason suggested.

'difference in infrastructure' Reason linked to graph.

'some rural regions the development has been slower' Reason expanded and linked to graph.

'well developed infrastructure for drinking water' Valid reason given.

(c) International agreements have a significant role to play in the promotion and protection of human rights. They outline the fundamental rights that all people are entitled to regardless of who they are. This includes freedom and equality, justice and security, wellbeing and education. Two major international agreements on human rights are the UN's Universal Declaration of Human Rights (UDHR) and the Council of Europe's European Convention on Human Rights (ECHR), both of which have been welcomed and criticised. International agreements such as these set out standards which can be followed by countries. They identify the key aspects of human rights and can be used to solve problems when people feel that their rights have been violated. They also offer the minimum protection to people in conflict areas, e.g. refugees. Some countries receive aid on condition that they meet human rights standards. However, such agreements have been criticised as some groups believe that they are 'too western' and do not account for cultural differences. For example, some in Islamic states have argued that the UDHR cannot be put into practice by Muslims without breaking Islamic law. This has led to the Cairo Declaration on Human Rights in Islam, which gives an Islamic perspective on the issue. Not all countries sign up to these agreements and they are difficult to enforce.

(e) **The student demonstrates a good understanding of the role of international organisations in the context of human rights. They explain briefly what agreements do, name examples of particular agreements and put forward some of the advantages and disadvantages of them. An example of a situation where the agreements have been successful would gain another mark. Level 3, 7 marks**

(d) Development aid is help given to developing countries. It is assistance that can be given in the form of money (grants and loans) and also equipment and expertise. There are different types of development aid such as emergency aid, NGO aid, bilateral and multilateral aid, and each approach has strengths and weaknesses. While aid can help reduce problems, it can also be less effective than intended and can actually lead to more issues.

Countries are given money by international organisations such as the World Bank and the IMF to help them improve their economies. However, this money comes in the form of loans on condition that a country restructures its economy or liberalises its trade. This may place the country at a disadvantage in the global trading system as TNCs can access markets more easily. This makes domestic firms less competitive. In Zambia, meeting the terms of the SAP led to reductions in the number of workers in the public sector which led to increased unemployment. Changes in healthcare provision made it more difficult for poorer families to access treatment. Cuts in education spending (from $60 to $15 per primary child) during the 1990s and the introduction of school fees negatively affected primary education.

Zambia's GDP per capita has risen sharply since the 1990s but remains well below the world average. However, multilateral aid can bring benefits. For example, the World Bank's Sri Lankan Pumpkin Tank project improved the water supply to those in remote areas at low cost.

Bilateral aid between governments of rich and poor countries can fund large infrastructure projects. For example, hydroelectric power schemes can bring a reliable, renewable energy source to the population. However, sometimes these projects are linked to 'tied aid' such as the Pergau Dam in Malaysia, where the UK government agreed to help fund the dam if the Malaysians bought arms from them.

NGOs can focus their efforts on a particular issue, often in a particular place. For example, Kibera in Need is a UK-based charity which helps people living in this Nairobi slum. It gives assistance to disadvantaged people by providing education, vocational and business skills. This provides long-term benefits as people can try and work their way out of poverty and has been successful. However, NGOs can only function if they receive donations from governments or the public. If these donations stop, this makes the people that they care for even more vulnerable. Emergency aid, such as MSF providing medical care and shelter following the 2012 earthquake in Haiti, can help to solve problems in the short term, but further strategies are needed in the medium and long term to improve living standards in the world's poorest countries.

One major criticism of development aid is that it can lead to dependency as poor countries remain reliant on outside funding and can be wasteful as the money spent may not be used properly or may be exposed to corruption. Smaller NGOs may offer a more successful model of aid but not reach many people. It is likely that future development strategies will include some form of aid but also promote foreign investment and fair trade in order to move towards a more sustainable model of development.

(e) **Good understanding of the problems of development aid is demonstrated. The student supports their points with examples from a range of types of aid and from different organisations. There is a good level of detail, which shows geographical knowledge. The conclusion is linked to some of the evidence in the essay but the connections to the different types of aid could be clearer.** **Level 4, 17 marks**

Student B

(a) (i)

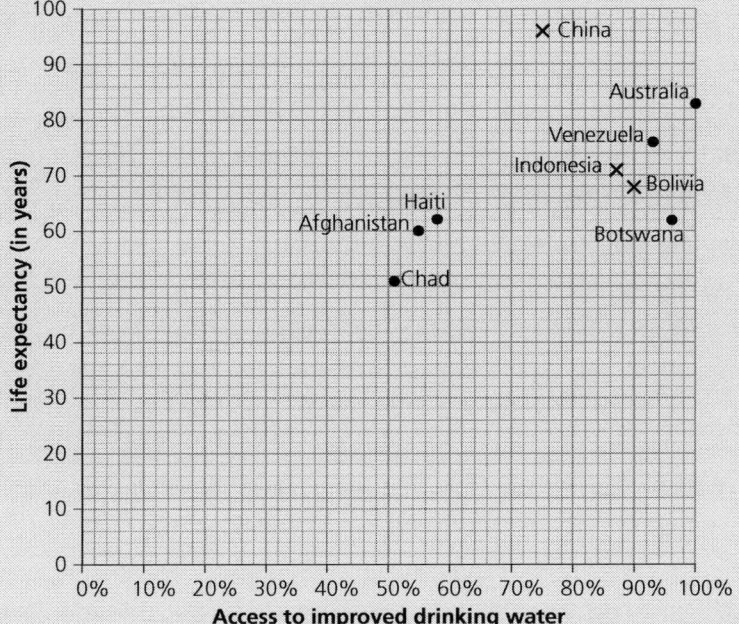

(e) The student has plotted Bolivia and Indonesia correctly but has misinterpreted the data for China. **2 marks**

(ii)

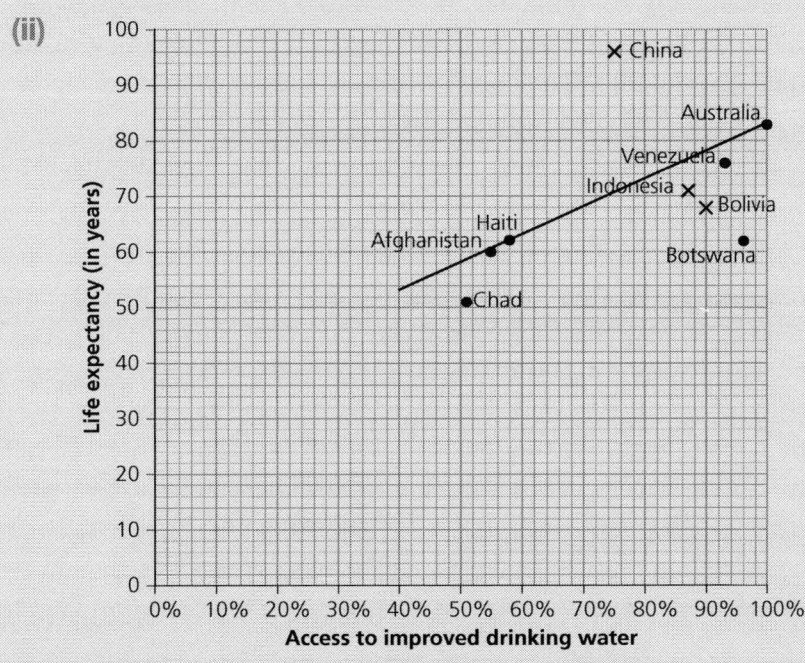

(e) The student has drawn a line but it is not accurate. **0 marks**

(b) There is a relationship between drinking water and life expectancy. This is because you need clean water to survive. If the water is bad then there may be diseases in it and this can make people unwell and they could die. Poorer countries find it difficult to supply all of their people with clean water because it is expensive to connect everyone to the treated water. The landscape might make it difficult to get the clean water to everyone and wars may damage pipes. Some countries which are not really developed like Botswana may have good access to water as the government may have spent money on its systems. However, other factors like HIV may affect the health of people and this affects life expectancy in a bad way.

(e) **The student does not state clearly what the relationship is. Valid reasons are given and the student attempts to group some countries together. The Botswana example is well used to clearly link the point to the graph. More detail is needed.** **Level 2, 4 marks**

'There is a relationship' Not creditable without stating what the relationship is.

'water is bad then there may be diseases in it' Connection made between variables.

'difficult to get the clean water to everyone and wars may damage pipes' Reason given but could be expanded to link to an example from the graph.

'spent money' Reason given and link to example from graph.

(c) The Universal Declaration on Human Rights is an important international agreement which focuses on ensuring that all humans have the right to freedom, justice and equality. It was set up by the United Nations in 1948. It has 30 articles which list different things that people are entitled to. It says that we are all equal and must not be discriminated against. We must be able to live securely and receive fair justice. People have freedom to move and are allowed to find safety. It is supported by Amnesty International. These rights belong to everybody and they can't be taken away. A government is not allowed to decide which parts of the agreement it likes and which parts it doesn't. The European Convention on Human Rights was signed by members of the Council of Europe and is also an agreement which protects freedom and human rights. If countries break the agreement, they have to go to the European Court of Justice to settle the problem. International agreements are important because they set out what everyone is entitled to and can force governments to change if their attitude to human rights is not good enough.

(e) **This answer shows some knowledge of international agreements but is focused around a basic description of the function of two agreements (UDHR and ECHR). The overall importance of such agreements is not discussed until the last sentence. Some advantages of such agreements are outlined, but the student could include the criticisms that some groups have of them.** **Level 2, 4 marks**

(d) Development aid aims to help poorer countries deal with their problems. Aid can be given to governments so that they can build infrastructure that will help their economy to grow, e.g. roads for transporting goods and dams to provide electricity. Aid can also be given to countries that need assistance following a natural disaster like an earthquake or a hurricane as people can be given food and shelter and helped to rebuild their lives. Charities can be effective at helping people in poor countries as often they are focused on a particular problem, such as providing malaria nets or educating children so that they have a better chance of getting a job. However, sometimes NGOs can be criticised. For example, Playpumps International in South Africa built roundabouts for children to play on. As the roundabout turned, water was pumped from the ground up to a water tank. The problem was that the pumps were too expensive and were difficult to maintain. There were also questions asked about using children as unpaid labour to supply a basic need for the community.

Top-down aid projects also bring problems as the money that is given to poor countries is controlled by the government. If the government is corrupt and not politically stable, then the money will not be used to help people. Also some governments only receive aid if they meet certain conditions. For example, Uganda and Zambia had to change how their economies worked as part of an SAP. This meant the governments spent less on healthcare and education. Instead of benefiting from the SAP, the populations have a lower standard of living and an increase in infant mortality and unemployment. Tied aid also means that a government has to buy a product, like weapons, from a richer country as part of a deal for aid money, like the Pergau Dam project.

It is true that development aid can cause problems for developing countries. Aid is supposed to help people through projects which improve their standard of living and increase their opportunities. Some projects are successful, such as the WHO's meningitis A vaccination project which has helped to control this deadly disease in Africa. However, certain conditions have to be met for some types of aid to be given, which may harm the country's economy and make it difficult for poorer people to benefit from it.

e **The student demonstrates some knowledge and understanding of the strengths and limitations of development aid. Some examples are given of several different types of aid but often the points are general and not supported well by the evidence. Ideas need to be linked to specific case study detail where appropriate; the Playpumps example links well to the question. A brief conclusion is made with some links to the points made in the essay.** **Level 3, 12 marks**

Question 4 mark scheme

(a) (i) 2 marks (AO3 = 2 marks)

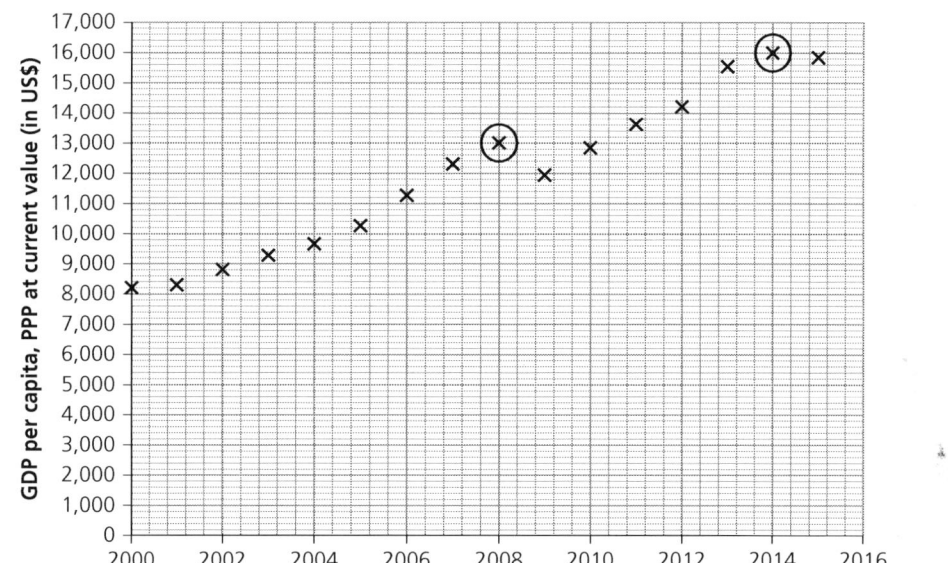

You gain 1 mark for each correctly positioned point. You must be accurate: there is no tolerance on either the horizontal axis (Year) or the vertical axis (GDP per capita, PPP at current value in US$).

(ii) 2 marks (AO3 = 2 marks)

You gain 1 mark for showing your working:

$$\frac{2015 \text{ figure} - 2000 \text{ figure}}{2000 \text{ figure}} \times 100$$

$$\frac{15,800 - 8,200}{8,200} \times 100$$

You gain 1 mark for the correct answer:

$$= 92.7\%$$

(b) 6 marks (AO1 = 3 marks, AO2 = 3 marks)

This question examines the factors that are considered when measuring human development. Measuring progress has been traditionally linked to GDP growth. However, a range of other economic, social and political components may be considered to be important too. Some suggested ideas are given below but you may wish to expand on these or include other relevant points.

AO1 Demonstrating your knowledge and understanding
➤ Human development can be measured in many different ways, not just by economic growth.
➤ Single variables can measure specific indicators which show levels of development.
➤ Composite variables can combine significant indicators to give a more overall view of development.

AO2 Applying your knowledge and understanding
➤ While it is important to consider GDP growth, other components (e.g. social, political) should be considered to reflect the complexity of measuring human development.
➤ The inclusion of different components may reflect varying viewpoints between different groups as to what human development means.

Answers to this question will be given a mark within a level band

Level 1 (1–2 marks) You show some geographical knowledge and understanding about the measurement of human development but some points are inaccurate. Your knowledge is not applied consistently with the question.

Level 2 (3–4 marks) You show mostly relevant geographical knowledge and understanding about the measurement of human development. Your knowledge in general is applied consistently with the question, although with only some details.

Level 3 (5–6 marks) You show accurate and relevant geographical knowledge and understanding about the measurement of human development. You apply your points logically to the question and show a good level of detail.

Hints and tips
Think about how human development can be measured in other ways than through things to do with money. Are there some aspects of human development that may be included by some groups? Are there some aspects of development that may be harder to measure than others?

(c) 8 marks (AO1 = 8 marks)

This question examines why levels of political corruption can vary between countries. Political corruption includes the ways in which elected officials can benefit illegitimately through their position and influence. Suggested ideas are outlined below, but you do not need to include all of these in your response. Other relevant points will be given credit.

AO1 Demonstrating your knowledge and understanding
➤ Political corruption is the way in which politicians can make illegitimate personal gain from their position.
➤ The level of political corruption varies between different countries for reasons such as lack of transparency, abuse of position, culture.
➤ Political corruption is difficult to measure but indices such as the Corruptions Perceptions Index have been developed to try to work out the extent of the problem.

Answers to this question will be given a mark within a level band

Level 1 (1–2 marks) You show limited knowledge and understanding of why political corruption varies between countries, and there are some inaccuracies. You lack a range of geographical ideas.

Level 2 (3–5 marks) You show some relevant geographical knowledge and understanding of why political corruption varies between countries, and have a range of geographical ideas. Your answer lacks detail.

Level 3 (6–8 marks) You show good geographical knowledge and understanding of why political corruption varies between countries. Your points are accurate, relevant and include a range of geographical ideas which you put forward in detail.

Hints and tips

What is political corruption? Why might some countries have higher levels of it than others?

(d) 20 marks (AO1 = 5 marks, AO2 = 15 marks)

This question focuses on evaluating the success of the UN's Millennium Development Goals (MDG) initiative. While there has been a more favourable outcome for some of the goals set, there has been less progress with others. In addition, some areas of the world have seen more improvements than others. Suggested ideas are outlined below, but you do not need to include all of these in your response. Other relevant points will be given credit.

AO1 Demonstrating your knowledge and understanding
➤ Levels of success can be assessed for each of the eight MDGs set and their targets.
➤ There has been mixed progress for the MDGs in terms of the targets set.
➤ Some individual countries and regions of the world have seen more progress than others as a result of the MDGs.

AO2 Applying your knowledge and understanding
➤ Despite the MDGs, there are still some areas of the world, e.g. sub-Saharan Africa, which face significant development challenges.
➤ Criteria for success vary widely between goals and were ambitious in some cases. This may affect the level at which targets can be achieved.
➤ While some goals, e.g. eradication of extreme poverty and gender equality, have shown much progress, there have been more inconsistent results with others, e.g. combating HIV.
➤ External factors, e.g. high population growth rate, may have had an impact on final outcomes.
➤ The post-2015 SDGs (Sustainable Development Goals) could be seen as showing the ongoing need for a global development initiative, reflecting the fact that the MDGs were not all met. Alternatively, the SDGs could be viewed as a continuation of an initiative that was successful, i.e. building on progress so far.

Answers to this question will be given a mark within a level band

Level 1 (1–5 marks) You show isolated points of knowledge and understanding about the level of global success of the MDGs with some errors and inaccuracies. You show limited understanding and are not always able to make connections between your points. Your answer is incoherent and lacks relevant evidence to support ideas. Your argument is limited, with unbalanced points. Your ideas are concluded in a general manner, if at all.

Level 2 (6–10 marks) You make some points showing knowledge and understanding about the level of global success of the MDGs, some of which may be relevant. You make some inaccurate points. You apply some knowledge but your points are not developed or may not be linked to the question. You use some evidence to support statements which may answer only part of the question. You draw a conclusion but this is based on unbalanced ideas.

Level 3 (11–15 marks) You show geographical knowledge and understanding about the level of global success of the MDGs. Your ideas are mostly relevant to the question and you make accurate points. You focus on at least two MDGs and/or countries/regions. You make some connections between ideas. You interpret the question well in general, but there may be some gaps in your use of evidence to support points. You draw a conclusion which links to the arguments made but is not fully supported by evidence.

Level 4 (16–20 marks) You show good use of geographical knowledge and understanding of the level of global success of the MDGs. You make a range of relevant points, focusing on at least three MDGs and/or countries/regions to create a coherent argument supported by relevant evidence. All your points are linked to the question. You draw a good, well-balanced conclusion which links clearly to the evidence presented.

Hints and tips

Have some of the eight MDGs been more successful than others? Can you discuss three MDGs in detail and say whether they have had positive outcomes?

Question 4 example responses

Student A

(a) (i)

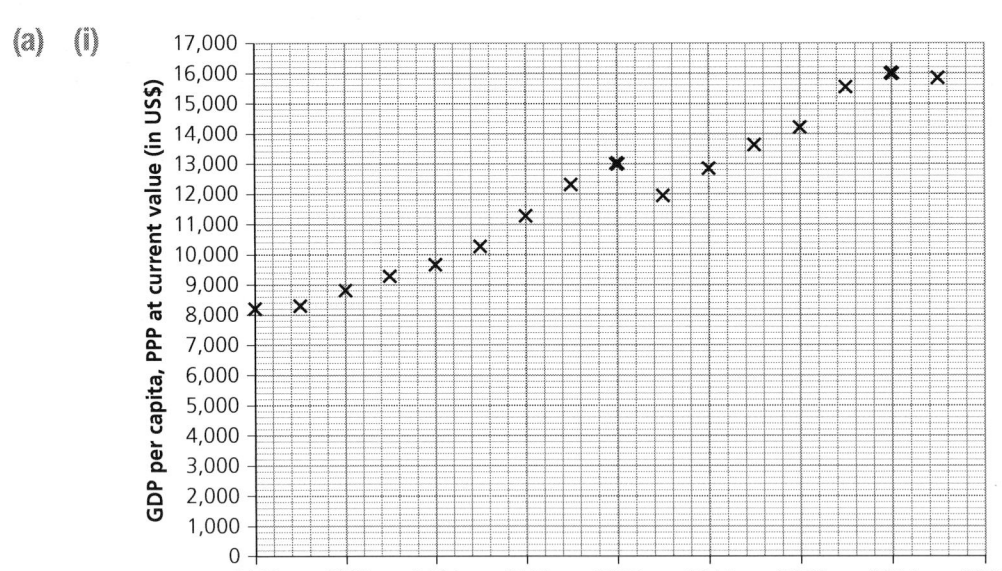

Ⓒ **Two correctly positioned points.**   **2 marks**

   (ii) $\dfrac{15{,}800 - 8{,}200}{8{,}200} \times 100 = 92.7\%$

Ⓒ **The student gives the correct answer and shows their working.**   **2 marks**

(b) Measuring human development is a <u>complex and contentious issue</u>, involving many aspects of life, e.g. health, education, freedom. Many variables are needed to try to do this effectively. Such measurements are made so that it is easier to compare different countries and so assess their 'level' of development. Traditionally, using just economic factors such as GDP growth has been seen as an effective way to establish which countries are making most financial progress in a given timeframe. However, it is not just the value of the money but the <u>effectiveness of how it is used</u> that is important. While variables linked to GDP play a significant part when assessing development, multivariate indicators such as the <u>Human Development Index</u> combine a broader range of data such as life expectancy at birth and years of schooling, as well as income per capita data. This helps to assess <u>health and education factors</u> which are important to a nation's progress. The Happy Planet Index includes measurements on wellbeing, inequalities of outcomes, and ecological footprint as it focuses on outcomes that can be <u>linked to sustainable development</u>, which includes environmental considerations. However, some aspects of development are hard to put a numerical value on (e.g. freedom of speech), which can affect the meaningfulness of the data collected.

ⓔ **The student gives reasons why measuring human development is not just linked to economic factors. They explain their points well and make good use of the HDI and HPI examples to expand their ideas.** **Level 3, 6 marks**

'complex and contentious issue' Reason given why many components are needed.

'effectiveness of how it is used' Limitations of using solely economic factors.

'Human Development Index' HDI example linked to question with detail.

'health and education factors' Link to specific aspects of development.

'linked to sustainable development' Example linked to question.

(c) Political corruption is the way in which elected officials in governments abuse their position for personal gain. Politicians can benefit from a range of illegitimate activities such as manipulating policies or procedures during large development projects. When contracts are handed out, some officials may take bribes in order to award the contract to a particular company. This behaviour is used by some officials to keep hold of their power, status and wealth. While it can happen at local, regional and national level, it can be difficult to detect and to change the culture which promotes such activities. The level of corruption can be difficult to measure precisely because of the hidden and often illegal nature of the activity. The Corruption Perceptions Index by Transparency International tries to do this by assessing the level of corruption in different countries. Some countries are very clean, e.g. Denmark, whereas others are highly corrupt, e.g. Somalia. Countries which are clean tend to be more transparent with their procedures and allow public access to information through the freedom of information acts. It may also be safer for whistleblowers to talk about corruption as they are less likely to be put in jail. Politicians themselves may have to be more accountable to the public, depending on the political system they belong to. Levels of political corruption between countries can also vary depending on the resources available. For example, politicians in countries with highly valuable natural resources, e.g. oil, or which are subject to large development projects, may have more opportunity to gain large amounts of money from decisions made. However, it is extremely difficult to measure the exact extent of corruption in a country, making comparisons between countries tricky.

ⓔ **The student shows good knowledge and understanding of political corruption and makes valid suggestions as to factors that might affect the level of corruption in different countries. Relevant points linked to the Corruption Perceptions Index are made. Further detail about corruption in one studied country would gain further marks.** **Level 3, 7 marks**

(d) The MDGs were a set of eight goals which were set by the UN in 2000 to address some of the key causes of poverty and underdevelopment in the world. Each goal focused on one key aspect, namely eradicating extreme poverty and hunger, achieving universal primary education, promoting gender equality, reducing child mortality, improving maternal health, combating disease (e.g. HIV), ensuring environmental sustainability and developing a global partnership for development. Measurable targets were set for each goal, to be achieved by 2015.

There has been varying success in the goals relating to health. Health issues were a major focus of the MDGs as many vulnerable people in the world's least developed countries are affected by problems which can be addressed more effectively by improved healthcare. For example, MDG 5 contributed to a 45% global reduction in the maternal mortality ratio. Although this did not meet the target set, progress in eastern Asia, northern Africa and southern Africa did achieve the 66% target. Projects, such as training full-time midwives in Bangladesh and Rwanda's RapidSMS initiative (where pregnancies can be tracked by health workers on mobile phones), contributed to this success. However, MDG 6 had more mixed success. While the number of new HIV cases decreased by 33%, the UN still estimates that over 2 million people are affected by HIV each year, with over 1.5 million of those being in sub-Saharan Africa. However, progress has been significant against TB, and malaria deaths have been reduced by initiatives such as the mosquito net campaign in DRC.

One major success of the MDG initiative has been targets linked to eradicating extreme poverty. Extreme poverty rates have fallen in every region and globally. The target of reducing the rates of those living on around $1 per day was achieved 5 years before the deadline. However, although the number of people in this situation has halved, there are still over 800 million people living in extreme poverty. MDG 3 has also seen much success. This MDG focused on gender equality. Giving girls access to education is seen as a key development criterion as it empowers more of the population. Southern Asia has seen the most progress with gender parity in primary schools and women's rights are more visible in some countries. However, in sub-Saharan Africa, the gender gap has widened in university education. In 2000 there were 66 girls per 100 at universities, and in 2011 this had dropped to 61 per 100. Poverty and entitlement issues still need to be addressed if this trend is to be reversed.

However, there have also been critics of the goals who suggest that there was a lack of justification behind their choice. Some of the targets are also difficult to measure. It is fair to say that, as with many wide-ranging initiatives, the MDGs had a varying range of success. The targets that were set were ambitious and external factors, such as increasing population growth in many poorer countries plus the economic crisis of 2008, may have had an impact on the MDGs' achievements. It could be argued that the MDGs have raised the global profile of some key development issues and have tried to turn this awareness into action and change. While overall there has been much progress made by the MDGs, there is still work to be done to improve the quality of life for the world's poor. Sub-Saharan Africa remains a region with poor quality of life for many, whereas parts of southern Asia have seen much progress. Such global disparity will need to be addressed further in the current UN Sustainable Development Goals initiative.

ⓔ **The student clearly understands the question and applies a range of specific case study knowledge to explain their points. The answer shows a balanced approach, as successes are highlighted but areas where outcomes have been less positive are also addressed. At least three MDGs are discussed — a requirement to reach Level 4. There is a good conclusion.**

Level 4, 17 marks

Student B

(a) (i)

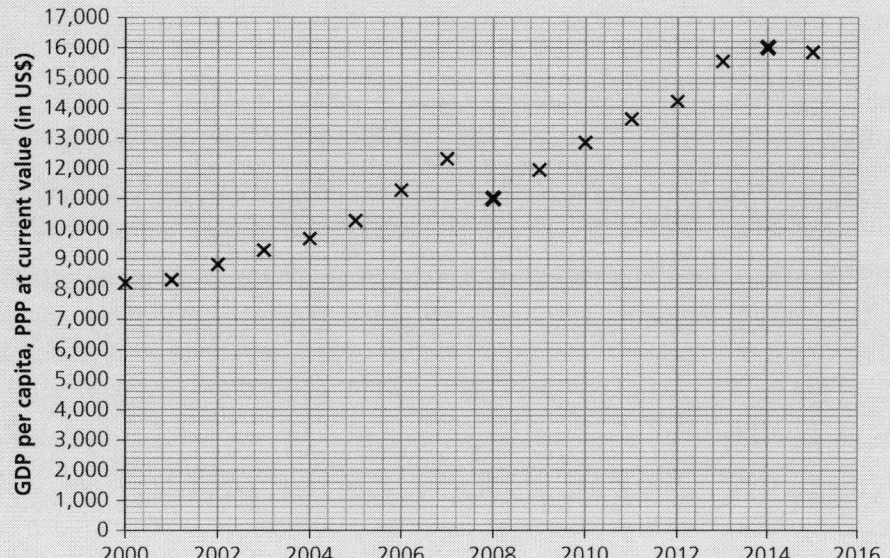

ℯ **Only one data point has been positioned correctly.**   **1 mark**

   (ii) $15{,}800 - 8{,}200 = 7{,}600$

$$\frac{7{,}600}{100} = 76\%$$

ℯ **Incorrect calculation using the data.**   **0 marks**

(b) It is difficult to measure the level of a country's human development because there are many factors which can influence this. Although a country can be seen to progress if its level of GDP increases, this money may not benefit the people if it is <u>not used in the correct way</u>. If you look at things like <u>life expectancy</u>, you can see whether people are generally healthy. This can be linked to whether they can afford <u>decent food and good healthcare facilities</u>. If GDP is high then there might be better investment in hospitals and so people have a better chance of staying in good health for longer. However, it should be noted that it is <u>difficult to collect up-to-date, accurate data</u> in some cases.

ℯ **The student shows good understanding of the importance of other components as well as the link between GDP and investing in services to promote development. However, the answer is brief and could include examples of development indices to help expand key points.**   **Level 2, 3 marks**

'not used in the correct way' Point linked to how money is used, rather than just its value.

'life expectancy' Reason why life expectancy is a good indicator of development.

'decent food and good healthcare facilities' Point extended with link to diet and healthcare.

'difficult to collect up-to-date, accurate data' While correct, not linked directly to the question.

(c) The amount of political corruption can be different in one country as opposed to another. Political corruption is when politicians abuse the system they work in. They often have to make decisions on big-money projects like dams and weapons. Sometimes the companies which stand to gain from the business may give a payment to the politician to help them make the 'right' choice. The company wins because it gains the contract and the politician wins because they get richer, but the country and the public lose out because this might not be the best deal. Whether a politician is likely to do these kinds of deals depends on the culture of the country. In a democracy, politicians are more likely to be accountable to their voters and have to justify the decisions that they make. They may have to show documents on the web and may have to have public meetings to explain what they are doing. Any business that takes place must have accounts and be audited, so any 'missing money' which might have gone to the politicians can be identified. However, in a dictatorship, the leader can do want they want and does not have to show the paperwork. People may also be fearful of any punishments that might take place if they speak out about corruption.

e **The student shows a clear understanding of corruption and discusses some of the factors that influence its level. Different levels of corruption depending on type of government are discussed, but no examples of countries that may have high or low levels of corruption are given. Suggesting types of projects within a country that may be more exposed to political corruption may gain further marks.** **Level 2, 4 marks**

(d) In 2000, the UN started the MDG initiative to tackle some of the world's most significant problems. Even in the twenty-first century, there are many people who do not have enough food to eat, do not have clean water and sanitation and who do not have basic rights. The UN decided that it would support different projects around the world to help get people out of poverty and to give them a better quality of life. There are eight goals and each goal has many targets which had to be met by 2015. Some goals were easy to measure but others were more difficult.

I am going to talk about two MDGs. First, Goal 2 aimed to have every boy and girl complete primary schooling by 2015. It is important that young children go to school as they learn how to read and write and also there is a better chance that they will go on to secondary education. Literacy rates have risen, girls have more chance to go to school and the number of children who do not go to primary school has decreased a lot. However, many students still do not complete school because their parents cannot afford to send them or children may have to stop school to earn money for their families in some places. I think that this MDG has been quite successful as 90% of children now go to primary school, but there are still big differences between the number of girls and boys in school in some countries.

Second, MDG 7 focused on environmental sustainability. This MDG has been successful in some ways. More than 2 billion people now have access to improved drinking water sources. Clean, affordable and accessible water is necessary as it is needed for people to live. If the water is not clean then diseases like cholera can spread and this can cause more problems and need to be treated. Although more people have better sanitation, there are still over 2 billion people who do not have a proper toilet. Many people also live in slums and shanty towns which have poor living conditions. Despite the MDG, many ecosystems are under threat. Although there have been more laws to protect the land, sea and atmosphere, there are still many species that will be extinct if their habitats are not saved. Deforestation is still continuing at a fast pace in some countries and this also affects the wildlife that lives there.

I think the MDGs have had some success but there is still a long way to go if people's lives are to be improved.

e **The student shows some knowledge and understanding of the goals. Two goals are discussed in more depth, although more detail about how the targets were met in different parts of the world or within different countries would benefit the answer. More and less successful aspects of the goals are discussed. There is a conclusion, but this is too brief and arguments are not drawn together.** **Level 3, 13 marks**

Option 8B Migration, identity and sovereignty

Question 5 mark scheme

(a) (i) 3 marks (AO3 = 3 marks)

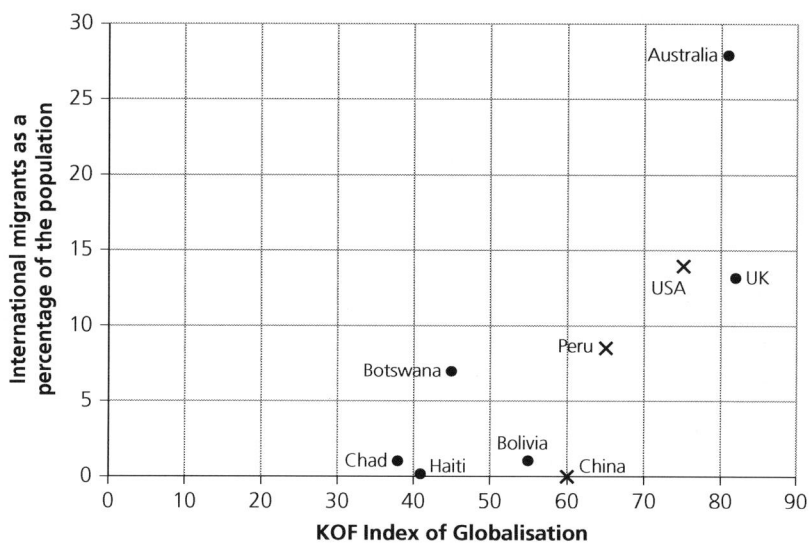

You gain 1 mark for each correctly plotted point for USA, China and Peru.

(ii) 1 mark (AO3 = 1 mark)

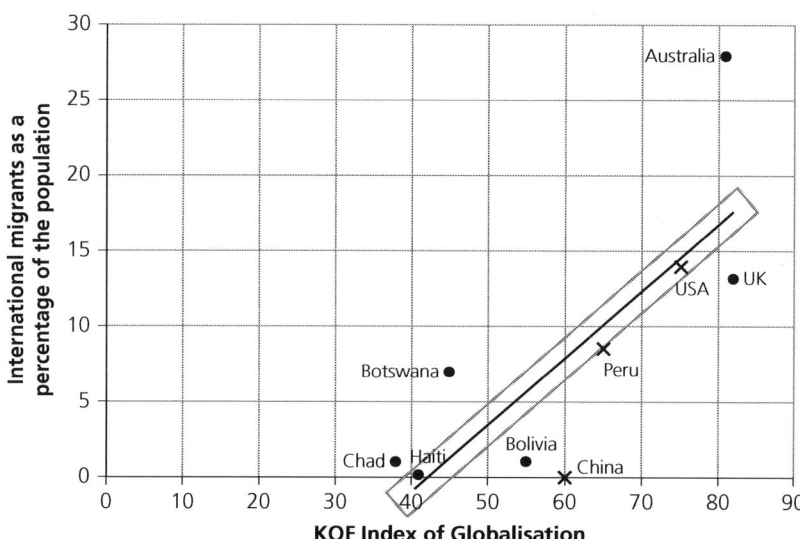

The best fit line for the graph is drawn in the centre of the rectangle. Parallel lines within the rectangle will be given a mark.

(b) 6 marks (AO1 = 3 marks, AO2 = 3 marks)

This question examines the relationship between international migrants as a percentage of the population and the KOF Index of Globalisation. You should recognise that there is a positive correlation between the variables, i.e. as the percentage of international migrants increases, so does the Index of Globalisation. Suggested content is below but you do not need to include all of it in your answer. Other relevant material can be credited.

AO1 Demonstrating your knowledge and understanding

➤ International migrants may be attracted to countries that are more globalised.

➤ More developed countries are likely to have an increased demand for a range of skills in different sectors of employment and international migrants may be able to provide these skills.

➤ Countries with lower KOF Index of Globalisation scores may be more protective towards their economies and have strict laws about the movement of workers.

➤ Countries with higher KOF Index of Globalisation scores may be more outward-looking and have a longer history of access to world markets.

AO2 Applying your knowledge and understanding

➤ Although the percentage of international migrants in general increases as the Index of Globalisation increases, there is a group of countries with low percentages of international migrants but a mid-level index of globalisation. This may be due to government policies restricting immigration.

➤ Countries with high levels of international migrants may be more open to the social aspects of globalisation, such as increased communication, the flow of ideas and people.

➤ Countries with high levels of international migrants may also have complex economies that are home to different parts of TNC operations. International migrants may find their skills are needed by such companies.

➤ Countries with high levels of international migrants may experience a high level of political globalisation. They can have a high diplomatic presence around the world and be members of a range of intergovernmental organisations.

Answers to this question will be given a mark within a level band

Level 1 (1–2 marks) You show some general knowledge and understanding of the relationship between proportions of international migrants and the KOF Index of Globalisation but some of this is inaccurate. Your knowledge is not applied consistently with the question. Some links you make between the graph and the question may be irrelevant.

Level 2 (3–4 marks) You show mostly relevant general knowledge and understanding of the relationship between proportions of international migrants and the KOF Index of Globalisation. You make some relevant links between the graph and the question.

Level 3 (5–6 marks) You show accurate and relevant knowledge and understanding of the relationship between proportions of international migrants and the KOF Index of Globalisation. You make logical connections between the graph and question.

Hints and tips

What is the main relationship between the variables? Can you group different countries? Link your reasons to examples from the graph.

(c) 8 marks (AO1 = 8 marks)

This question examines the idea that some borders that exist between countries can be disputed. There are historical reasons why these disagreements exist and different groups may view borders in different ways. Suggested ideas are outlined below, but you do not need to include all of these points in your response. Other relevant points will be given credit.

AO1 Demonstrating your knowledge and understanding

➤ National borders show the boundary between two nation states. They can be created by physical geography such as seas and rivers, or by historical agreements.

➤ Sovereignty is the idea that a state or government has the power to control the area over which it presides.

➤ Problems of sovereignty can include lack of recognition of the state by international law, processes that may not be completely controllable by the state itself (e.g. globalisation, international human rights,) the role of TNCs, cultural and ethnic differences, and the issue of failed states.

➤ National borders can be drawn up after colonialism and conflicts (e.g. Iraq), and may not take into account different linguistic, ethnic or religious groups.

Answers to this question will be given a mark within a level band

Level 1 (1–2 marks) You show limited geographical knowledge and understanding of why some national borders can lead to problems of sovereignty, with some inaccuracies. You show a limited range of geographical ideas.

Level 2 (3–5 marks) You show some relevant geographical knowledge and understanding of why some national borders can lead to problems of sovereignty. You use a range of geographical ideas but your answer lacks detail.

Level 3 (6–8 marks) You show a good geographical knowledge and understanding of why some national borders can lead to problems of sovereignty. Your points are accurate, relevant and include a range of geographical ideas, which are put forward in detail.

Hints and tips

Think about the notion of sovereignty. What can lead to disagreements of border areas? Can you give examples of borders which are disputed? Are there examples where borders become irrelevant?

(d) 20 marks (AO1 = 5 marks, AO2 = 15 marks)

This question focuses on the success IGOs have had when dealing with environmental problems. There have been a large number of initiatives put forward and supported by IGOs connected with atmospheric concerns, protection of species and their environments, and hydrological issues. IGOs have also been involved with continental management, for example in Antarctica. Different management strategies have had different levels of success. Suggested ideas are outlined below, but you do not need to include all of these in your response. Other relevant points will be given credit.

AO1 Demonstrating your knowledge and understanding

➤ There is a variety of management strategies IGOs can use to tackle global environmental issues.

➤ IGO management strategies can cover concerns in the atmosphere, the biosphere, the geosphere and the hydrosphere.

➤ These management strategies can have degrees of success depending on a range of factors, such as the extent to which they are legally binding, the motivations of the groups involved, how the strategy is implemented and how it is monitored.

AO2 Applying your knowledge and understanding

➤ IGO involvement may be necessary for recognition of global environmental problems and to provide a framework for management. Sometimes this is difficult to regulate and enforce.

➤ IGOs that are involved with managing the atmosphere include the Montreal Protocol on Substances that Deplete the Ozone Layer, which has resulted in the levelling or reduction of the concentration of CFCs and the protection of the ozone layer.

➤ IGOs that are involved with managing the biosphere include the Convention on International Trade in Endangered Species of Wild Fauna and Fauna (CITES) and the RAMSAR Convention on Wetlands.

➤ The management strategies of IGOs are a highly important way to protect marine environments. IGOs have been responsible for developing laws to protect and regulate the use of marine (UN Convention on the Law of the Sea) and fluvial (Water Convention) environments.

➤ Other environments such as international waters, transboundary river systems and Antarctica may benefit from an IGO perspective.

Answers to this question will be given a mark within a level band

Level 1 (1–5 marks) You show isolated geographical knowledge and understanding of the level of success of IGOs' management of environmental problems, with some errors and inaccuracies. You show limited understanding and you are not able to make connections. Your answer is incoherent and lacks relevant evidence to support ideas. Your argument is limited, with unbalanced points. Your ideas are concluded in a general manner, if at all.

Level 2 (6–10 marks) You show knowledge and understanding of the level of success of IGOs' management of environmental problems, some of which may be relevant. You make some inaccurate points. You apply some knowledge, but your ideas are not developed or may not be linked to the question. You use some evidence to support statements which may answer only part of the question. You make a conclusion but this is drawn from often unbalanced ideas.

Level 3 (11–15 marks) You show geographical knowledge and understanding of the level of success of IGOs' management of environmental problems. Your ideas are mostly relevant to the question and you make accurate points. You make some connections between ideas. You interpret the question well in general but there may be some gaps in the use of evidence to support your points. You draw a conclusion which links to the arguments you make but may not be not fully supported by evidence.

Level 4 (16–20 marks) You show good use of geographical knowledge and understanding of the level of success of IGOs' management of environmental problems. You make a range of relevant points to create a coherent argument supported by relevant evidence. All of your points are linked to the question. You make a good, well-balanced conclusion which links clearly to the evidence presented.

Hints and tips

Think about IGOs involved in tackling global environmental issues. What are the strengths and weaknesses of their management strategies? Can you give examples of successful and less successful management strategies? Can you find examples for different types of problems, e.g. atmospheric, ecological?

Question 5 example responses

Student A

(a) (i)

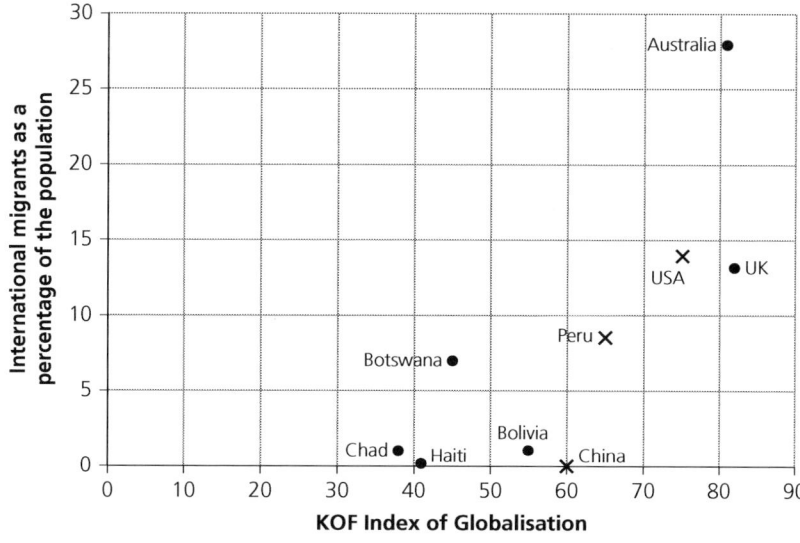

(e) Correctly plotted points for USA, China and Peru.   **3 marks**

(ii)

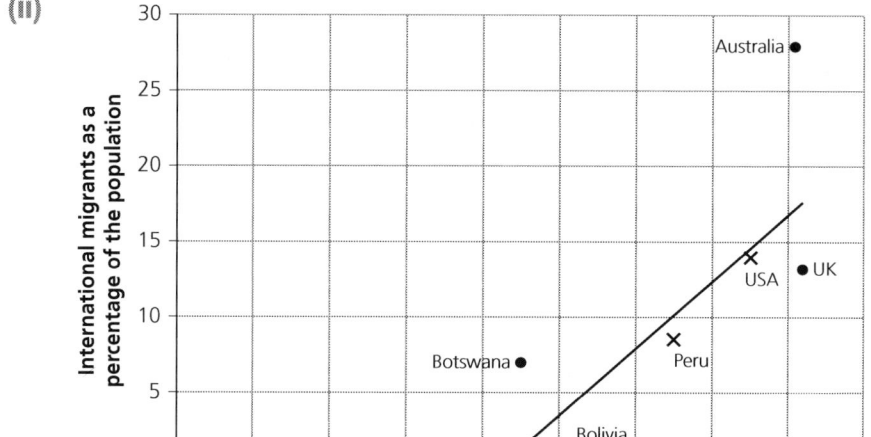

(e) The student has drawn the regression line accurately.   **1 mark**

(b) The graph indicates that there is a <u>positive correlation</u> between the percentage of international migrants and the KOF Index, although analysis of a <u>larger data set would be needed</u> to confirm this relationship. Countries such as Australia, the UK and the USA have higher levels of international migrants and high levels of globalisation. Australia, <u>with 28% of its population coming from abroad, has a tradition</u> of encouraging international migrants to meet skills shortages in the economy. The USA and UK are developed countries which have a long history of global trade and also of international migration. Migrants, who are needed to contribute to economic growth, may be attracted by existing ethnic communities and may create a more cosmopolitan society. On the other hand, countries such as <u>Chad and Haiti may have more emigration than immigration</u> and have fewer aspects of a globalised economy, e.g. low numbers of TNCs, fewer established mobile phone networks. China is a slight anomaly as it has a low percentage of international migrants (0.1%) but a medium level of globalisation. This may be because China has a large working population and does not need to attract migrants from elsewhere. The government may also have policies which control the number of foreign workers there.

(e) **This is a good answer that suggests valid reasons for the relationship but acknowledges their limitations. The student integrates data from the graph to support these reasons. There is an attempt to link country groups together and also to suggest an anomaly to the general pattern.**

Level 3, 6 marks

'positive correlation' Relationship suggested.

'larger data set would be needed' Limitations recognised.

'with 28% of its population coming from abroad, has a tradition' Valid reason given with clear links to data from graph.

'Chad and Haiti may have more emigration than immigration' Valid reason given with example from graph.

'China is a slight anomaly' Understanding of possible anomaly.

'China has a large working population' Valid reason for anomaly given.

(c) Sovereignty is a complex issue whereby a state or government has total authority within the boundary of a particular area. For many nations these borders are recognised by international law, but in many cases they are disputed by groups within the country as well as by other countries and international organisations. For example, the foundations for many borders in the Middle East were created after the First World War. The partition of territory by foreign powers such as the UK and France has led to a range of ongoing disputes over sovereignty in Iraq. Kurdistan is a region which extends into Iraq as well as Turkey, Iran and Syria. Groups seeking Kurdish independence wish to make Kurdistan a nation state and have its borders internationally recognised. Some countries also believe that they should have control over parts of other countries. For example, Gibraltar is a British overseas territory, having been captured and given to Britain in the early eighteenth century. However, its sovereignty is disputed by neighbouring Spain. Also, while the Republic of Cyprus has sovereignty over the island of Cyprus, the northern half has been under Turkish control since the 1970s, a situation which is seen as being illegal under international law. International organisations such as the European Union can also affect the role of borders in terms of a country's sovereignty, e.g. its Schengen zone allows free movement of people across borders without passport and custom controls. This may lead some groups to feel that this reduces a government's control over its territory and therefore threatens national sovereignty.

(e) **The student explains well the difficulties that national borders can cause in terms of sovereignty. They discuss boundary disputes, ownership of territory and loss of control. The student uses a broad range of examples (Iraq, Gibraltar, Cyprus and the EU) to support the points made, showing good geographical knowledge and understanding in the context of the question.** **Level 3, 7 marks**

(d) IGOs are organisations which are composed of different countries or other international bodies to address an issue which is of global importance. IGOs can tackle issues which may be difficult for individual nation states to solve, or that need a global approach to find a solution. While IGOs have the power to raise awareness of a global issue, to provide more information about the problem and to offer management strategies to help reduce damage caused, they do not always deal with environmental problems successfully.

Some IGO management strategies have been relatively good at achieving their aims. The Montreal Protocol is one of the most successful IGO management strategies. A high level of international cooperation and commitment through laws to ban the production and use of CFCs, HCFCs and halon has contributed to the protection of the ozone layer and has reduced the rate of its depletion. One of the reasons for its success was because there was a specific set of chemicals involved which could easily be banned.

The Antarctic Treaty System is an example of intergovernmental cooperation to protect one of the world's remaining wilderness regions. It has been successful at keeping the continent as a place for peace and science. This is because the Antarctic Treaty is relatively short and so can be adapted to new environmental threats. Many countries have signed the agreement, not just those with a claim on Antarctica. It has made it clear that certain activities such as mining and nuclear testing are prohibited. Also the decision-making body works by consensus so anyone can veto changes. However, the system may need to address the specific environmental issues posed by a rapid increase in tourism.

However, there has been limited success for other initiatives put forward by IGOs. For example, in its 2015 Millennium Development Goals the UN aimed to reduce biodiversity loss. While there has been an increase in the size of areas that have designated protection, some areas have inadequate management strategies that reduce their effectiveness. The Important Bird Areas only have 26% of their sites fully protected. Also, the Red List Index suggests that in general the number of species worldwide is decreasing. CITES has focused on reducing the trade of different species. While it has led to a reduction in the trading of ivory, it has been criticised for its focus on particular species which may grab the public's attention, rather than a more general approach to habitat protection and biodiversity.

IGOs can have their influence reduced if not all countries involved in the issue recognise the IGO or sign up to the recommendations that it makes. While IGOs may put forward solutions, they sometimes do not have a robust legal framework to implement their management strategies or to enforce their policies if countries disregard them. However, by increasing awareness of a particular issue, they are in a much better position than countries themselves or NGOs to highlight the importance of the problem on a global level. The physical processes and wildlife that exist in the Earth's atmosphere, biosphere and hydrosphere are beyond a single country's borders and are best protected by a global management approach.

The student addresses the question well. The general strengths and limitations of IGO management strategies are outlined and the student uses a range of examples (Montreal Protocol, Antarctic Treaty System, relevant MDG and CITES) to produce a balanced evaluation of IGO effectiveness. Some key points are summarised in the conclusion but reference could be made more clearly to the examples given in the main body of the essay. Level 4, 18 marks

Student B

(a) (i)

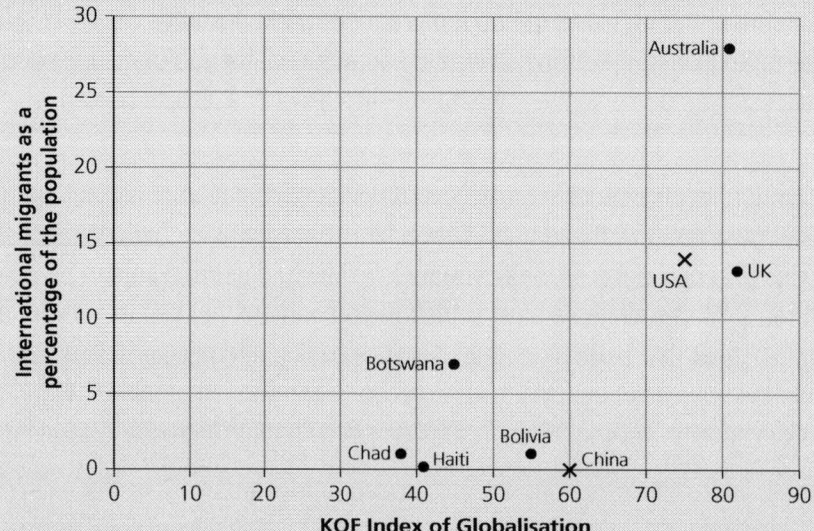

e The student has correctly plotted USA and China but has not plotted Peru. **2 marks**

(ii)

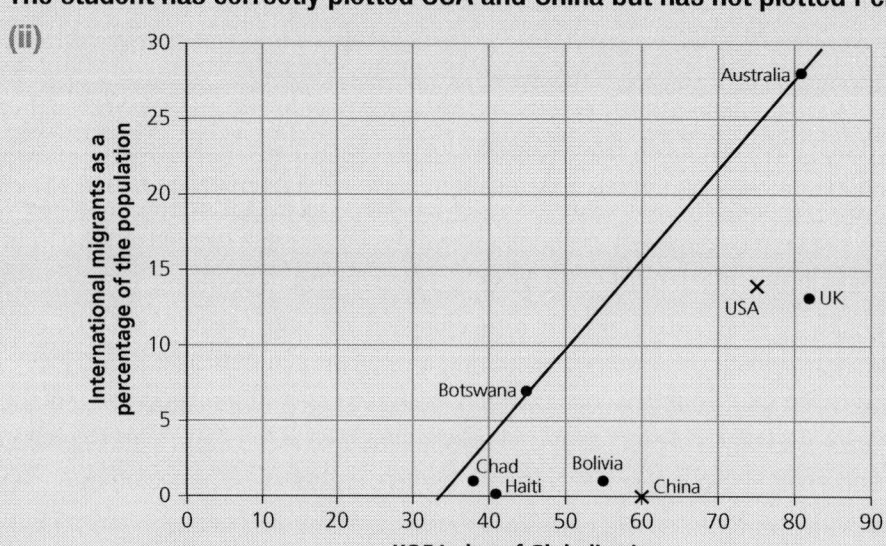

e The student has not drawn the best fit line correctly. **0 marks**

(b) Some countries have a large number of international migrants as <u>they can find a job</u>. It might be difficult to do this in their home country. They can <u>keep in touch with family and friends</u> through mobile phones and Facebook. This links them to other parts of the world and they can send money to them. <u>Haiti</u> is a poor country which suffers from a lot of natural disasters like earthquakes and volcanoes. International businesses do not invest there and people are more likely to leave there. This means that it is not so connected and so it has a low globalisation index. The <u>UK is well connected to global trade</u> and many workers from abroad work in TNCs in London as they have the right skills.

e **The student shows a basic understanding of the graph but does not really suggest a relationship between the variables. Some general valid reasons are given, and Haiti and the UK are used as supporting examples. Some understanding is shown, but more detail could be given to help justify the points made.** **Level 2, 3 marks**

'they can find a job' Some understanding of economic globalisation, but not clearly linked to the data.

'keep in touch with family and friends' Some understanding of social globalisation, but not clearly linked to the data.

'Haiti' Use of Haiti data to suggest valid reason.

'UK is well connected to global trade' Valid reason given, with the UK used to support.

(c) National borders show the limit of a government's control of its country. However, they can cause problems with sovereignty because neighbouring countries may not agree with where the border lies and try to take land back. This will cause conflicts to happen. Governments might not be able to control everything that they would like to within their country because of the internet. This may spread ideas that the government does not like. Sometimes international organisations may have rules that a country accepts when it joins the organisation which may also lead to a loss of power and control. For example, if companies want to move goods in Europe, they can move them across the border without being checked. Some groups of people within a country might not be happy about the location of the border, especially after a war. They might feel that their families might be split up or that their rights might not be recognised. Problems like this have happened in North and South Korea.

(e) **The student shows some geographical knowledge and understanding of what borders are, and puts forward some reasons why they might cause problems for a country's sovereignty. The role of international organisations is acknowledged. The example of moving goods around 'Europe' is not accurate and should have been linked more closely to the European Union. The Korean example is relevant, but a more detailed point linked to sovereignty would show a greater level of understanding.** **Level 3, 4 marks**

(d) Global environmental problems are difficult to solve. Air pollution is an example of an environmental problem which affects everybody. These issues may happen over a large area and may span many different countries. It is difficult for each country to solve the problem itself and so it is good to have an international organisation that is responsible for solving it. However, IGOs are not always able to find a solution to the problem. This is because it is difficult to get everyone to agree what action they should take. Also some countries may not sign up to join the IGO. In addition some groups may ignore the agreements and laws made by the IGO. For example, CITES is an IGO which has banned the trade in endangered species. However, there are still some people who illegally trade dead animals, e.g. parts of tigers for medicines.

Acid rain is a global environmental problem as one country can pollute the atmosphere but the winds blow it to another country, causing it a problem as its trees die. Some IGOs have been successful. The RAMSAR Convention protects wetland habitats across the globe. Hinuma Lake in Japan has been protected by reducing the number of days that the fishermen can fish for clams. This protects the wetland but also allows the local community to make a living. The Madrid Protocol has been successful. It banned the use of CFCs and now the hole in the ozone layer over Antarctica has been reduced. Environmental problems in the seas are difficult to manage. There is a lot of plastic which is thrown into the water by ships and it is difficult to find out who is to blame for this. Some charities like Greenpeace help to raise awareness of the problems that species can have when they eat rubbish in the sea. Finally the Water Convention is an example of an IGO which protects rivers that flow through different countries. It helps countries to cooperate so that they can all use the water from the river. This can be a problem if a country builds a dam and the flow of water is reduced in another country further downstream.

All in all IGOs have been successful in stopping some of the world's environmental problems. However, they have limited power in some areas and this makes it difficult for their rules to be kept in the actual place that needs protection.

ⓔ **The student understands the basics of the question and their answer gives examples used in context (e.g. Water Convention), although not always accurately (e.g. confusion between the Madrid and Montreal protocols; Greenpeace). The point on acid rain is not linked clearly to an IGO. The RAMSAR example is used more effectively. A general, valid conclusion is drawn, with some evaluation of IGO strengths and weaknesses, but this is not linked well to the evidence presented.** **Level 3, 11 marks**

Question 6 mark scheme

(a) **(i)** 2 marks (AO3 = 2 marks)

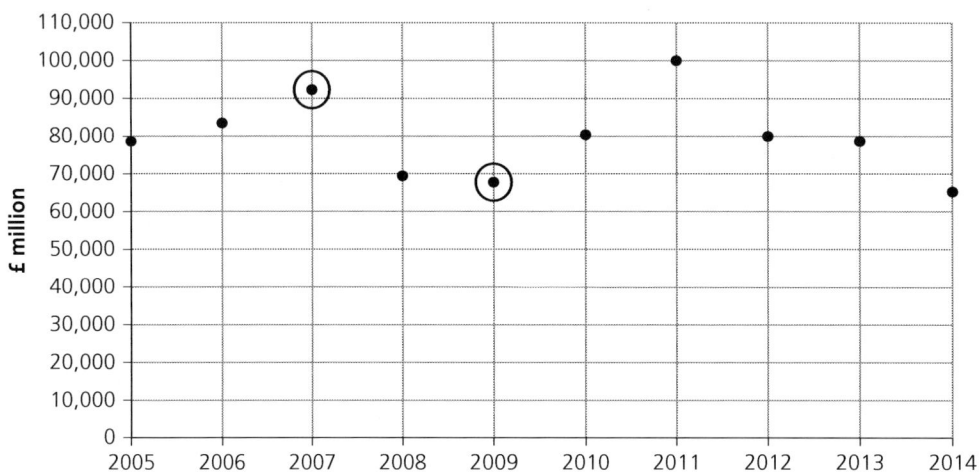

You gain 1 mark for each correctly positioned point. You must be accurate — there is no tolerance allowed on the horizontal axis (Year); however, a tolerance of up to 2,000 each way on the vertical axis (£million) is allowed.

(ii) 2 marks (AO3 = 2 marks)

You gain 1 mark for showing your working. An answer for the 2005 figure in the range 77,000 to 79,000 will be given credit. An answer for the 2014 figure in the range 64,000 to 65,000 will be given credit.

$$\frac{2014 \text{ figure} - 2005 \text{ figure}}{2005 \text{ figure}} \times 100$$

$$\frac{65,000 - 78,000}{65,000} \times 100$$

You gain 1 mark for the correct answer.

$$= -20\%$$

(b) 6 marks (AO1 = 3 marks, AO2 = 3 marks)

This question examines why foreign ownership of business and property can have an impact on national identity. Increasing levels of purchasing and possession of buildings, as well as local and national businesses, by foreign firms can lead to discussions of influence on national identity. Some suggested ideas are given below but you may wish to expand on these or include other relevant points.

AO1 Demonstrating your knowledge and understanding

➤ National identity is used to measure one's own affiliation to a particular country. It points towards the nation as a whole entity which has particular shared characteristics, e.g. traditions, culture and language.

➤ Foreign ownership of business and property is increasing in some economic sectors and in some places within countries. This can be linked to globalisation.

➤ Foreign ownership can raise issues about power, influence, control and security.

AO2 Applying your knowledge and understanding

➤ National businesses may feel under threat by foreign-owned or part-owned businesses in terms of how they run their business and where profits go to.

➤ Increased foreign ownership can affect the ability of people to buy property, either commercial or residential, as prices may be driven up. Increased foreign ownership of property can also bring opportunities for regeneration. This can have an impact on national identity.

Answers to this question will be given a mark within a level band

Level 1 (1–2 marks) You show some geographical knowledge and understanding about the impact of foreign ownership on national identity but some points are inaccurate. Your knowledge is not applied consistently with the question.

Level 2 (3–4 marks) You show mostly relevant geographical knowledge and understanding about the impact of foreign ownership on national identity. Your knowledge in general is applied consistently with the question, although with only some details. You may have included impacts from only business or only property.

Level 3 (5–6 marks) You show accurate and relevant geographical knowledge and understanding about the impact of foreign ownership on national identity. You have included impacts from both business and property. You apply your points logically to the question and show a good level of detail.

Hints and tips

Think about how foreign ownership of property can affect cultural traditions, e.g. home ownership in the UK. How does foreign ownership of companies have an impact on national identity?

....

(c) 8 marks (AO1 = 8 marks)

This question examines why nationalism in the nineteenth century was important in the development of empires. The development of nation states in Europe and their quest for resources and trade links played a key role in empire building. Suggested ideas are outlined below, but you do not need to include all of these in your response. Other relevant points will be given credit.

AO1 Demonstrating your knowledge and understanding

➤ Nationalism is the concept of loyalty and devotion to a person's own nation above all others.

➤ Nationalism in the nineteenth century links to the development of nation states, e.g. in Europe.

> ➤ After gaining their own independence in the nineteenth century, some countries looked to expand their influence and territory abroad.

> ➤ Expansionism in the nineteenth century can be linked to the need for natural resources and the exertion of power.

Answers to this question will be given a mark within a level band

Level 1 (1–2 marks) You show limited knowledge and understanding of why nationalism in the nineteenth century was important in the development of empires, and there are some inaccuracies in your answer. Your answer lacks a range of geographical ideas.

Level 2 (3–5 marks) You show some relevant geographical knowledge and understanding of why nationalism in the nineteenth century was important in the development of empires and you demonstrate a range of geographical ideas, but your answer lacks detail.

Level 3 (6–8 marks) You show good geographical knowledge and understanding of why nationalism in the nineteenth century was important in the development of empires. Your points are accurate and relevant, and you include a range of geographical ideas which are developed in detail.

Hints and tips

What is nationalism? Why did rulers of independent states feel they needed to expand their territory and/or influence in the nineteenth century?

....

(d) 20 marks (AO1 = 5 marks, AO2 = 15 marks)

This question focuses on evaluating whether highly skilled workers have more opportunity to migrate across international borders. The ability to migrate across international borders can also be linked to migrant income as well as migration policies set by governments at national and international level. Legal and illegal migration can also be considered. Suggested ideas are outlined below but you do not need to include all of these in your response. Other relevant points will be given credit.

AO1 Demonstrating your knowledge and understanding

> ➤ The migration of workers across international borders is connected to many factors, e.g. skill level, income, age, demand, government policy.

> ➤ There may be certain skills which are in higher demand than others.

> ➤ While some areas have free movement of labour, other areas may have more restrictions.

AO2 Applying your knowledge and understanding

> ➤ People with high skill levels are more likely to migrate across international borders.

> ➤ Migrants can be legal or illegal and this may affect their ability to move.

> ➤ Different levels of restrictions may affect workers with different levels of skills across the same international border.

Answers to this question will be given a mark within a level band

Level 1 (1–5 marks) You show isolated points of knowledge and understanding about the connection between highly skilled migrants and their ability to cross international borders, with some errors and inaccuracies. You show limited understanding and are not always able to make connections between your points. Your answer is incoherent and lacks relevant evidence to support ideas. Your argument is limited, with unbalanced points. Your ideas are concluded in a general manner, if at all.

Level 2 (6–10 marks) You make some points showing knowledge and understanding about the relationship between highly skilled migrants and their ability to cross international borders, some of which may be relevant. You make some inaccurate points. You apply some knowledge but your points are not developed or may not be linked to the question. You use some evidence to support statements which may answer only part of the question. You make a conclusion but this is drawn from unbalanced ideas.

Level 3 (11–15 marks) You show geographical knowledge and understanding about the relationship between highly skilled migrants and their ability to cross international borders. Your ideas are mostly relevant to the question and you make accurate points. You also put forward other factors which may influence the ability of different groups to cross borders. You use at least one example to show your points. You interpret the question well in general but there may be some gaps in the use of evidence to support points. You draw a conclusion which links to the arguments made but is not fully supported by evidence.

Level 4 (16–20 marks) You show good use of knowledge and understanding of the relationship between highly skilled migrants and their ability to cross international borders. You make a range of relevant points and use at least two examples to demonstrate your ideas. You also put forward other factors which may influence the ability of different groups to cross borders. All your points are linked to the question and you try to balance your ideas. You draw a good, well-balanced conclusion which links clearly to the evidence presented.

Hints and tips

Why are some jobs more in demand in some countries than others? What other factors affect whether people can migrate from one country to another?

Question 6 example responses

Student A

(a) (i)

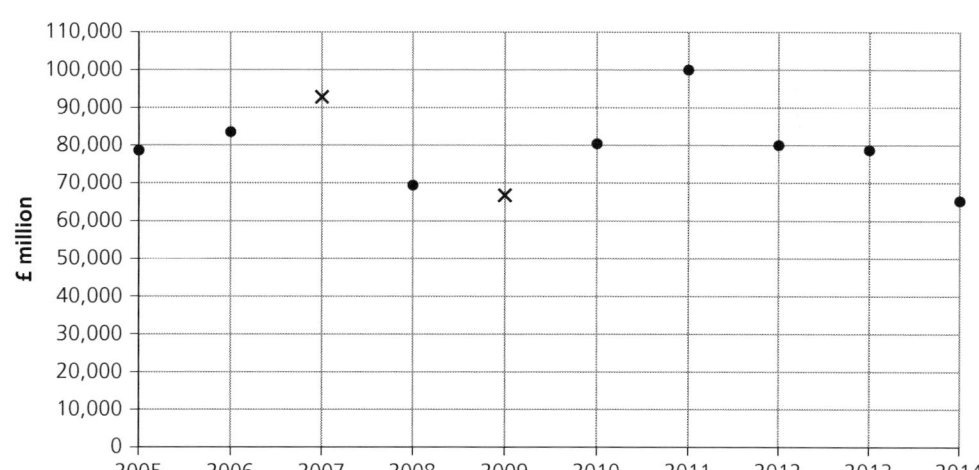

ⓔ Two correctly positioned points.   **2 marks**

(ii) $\dfrac{65{,}000 - 78{,}000}{65{,}000} \times 100 = -20\%$

ⓔ The correct answer is given, and working is shown.   **2 marks**

(b) National identity is the way in which a country is viewed by its people or how a person has a sense of belonging to one nation. As the world becomes more globalised, the likelihood that companies have <u>business interests outside their own country</u> increases and this can influence the culture of a country. Foreign companies which buy control of UK companies can cause uncertainty when they have a global review of their businesses. For example, jobs in the UK steel industry were under threat as <u>Tata</u>, an Indian multinational, reviewed its investment in its UK steel operations. This raises questions about national identity as the steel industry is not owned by the UK government anymore and so changes can come relatively quickly out of the government's control and have a direct impact on things that affect the people's <u>way of life, e.g. employment</u>.

In addition, increases in foreign ownership of property can have a large influence on national identity. Traditionally, the concept of <u>homeownership is very strong in the UK</u>, whereby people have a mortgage and eventually own their own home. However, in some parts of London, foreign ownership of large property blocks has increased. This can contribute to the rise in property prices in the capital, causing <u>lack of affordability for homeownership and a rise in renting</u>. Also many of these owners <u>do not use these homes as their permanent residence</u>, so key residential space is left empty rather than being lived in. This may change the feelings of community in a neighbourhood.

ⓔ The student answers the question well. Two key areas where foreign ownership can have an impact on national identity (business and property) are mentioned and relevant examples are given to explain the points made. **Level 3, 6 marks**

'business interests outside their own country' Foreign ownership linked to national identity.

'Tata' Example of influence of multinationals.

'way of life, e.g. employment' Link to job security and national identity.

'homeownership is very strong in the UK' Link to homeownership and national identity.

'lack of affordability for homeownership and a rise in renting' Link to homeownership traditions and national identity.

'do not use these homes as their permanent residence' Link to property ownership and change in communities.

(c) The nineteenth century saw a rise in nationalism, whereby nation states in Europe gained independence from former empires, e.g. Greece and the Ottoman Empire, or where regional states joined together to form a united country, e.g. Italy. Some countries then developed their own empires so they could access the natural resources that they needed for development and also exert regional power in areas that were further away, e.g. Southeast Asia. Some countries were ruled directly by the European state or were ruled by local leaders who were under the control of or sympathetic to the European colonists. Some countries such as the Belgian Congo were seen as the Belgian King's personal fiefdom.

For example, in the nineteenth century French nationalism grew following the defeat of Napoleon and military action was taken to expand the empire in North Africa, Indochina and the South Pacific. The new French colonies were able to supply raw materials and to also buy products that had been made in France. Thus, the colonies had an important trading role and also offered a strategic show of power in regions where other European nations were trying to expand their territories, e.g. Germany in Africa. Nationalism in Europe saw a rise in patriotism and the promotion of language, culture, education and government. The culture of the imperial nation was then spread to the new colony. For example, French nationalism promoted the spread of French and the French education system to its colonies; this still has a significant influence today, e.g. Morocco and Algeria. It can be said that nationalism had a strong link to the development of empires in the 1800s.

(e) **The student clearly links the concept of nationalism and its rise in nineteenth-century Europe to the growth of empires. Good understanding of key ideas is shown and points are expanded using the case study of France. The student could also mention why some regions seemed more susceptible to European colonists (e.g. because of existing diplomatic and trade links, proximity).**

Level 3, 7 marks

(d) The movement of highly skilled workers across international borders is complex. Highly skilled workers can refer to highly qualified staff with specific skills in sectors such as science, technology, medicine, engineering and IT. It also includes managerial and leadership abilities. The international migration of this type of workforce is increasing because of globalisation. As companies develop their businesses abroad, they want to recruit the best staff from the largest talent pool possible. Where skills are specialised, in great demand or where there is a shortage in a particular expertise, there may not be suitable workers in one country and so businesses will want to be able to attract staff across international borders to fill posts.

There is an increasing number of skilled workers, especially from Asian countries such as India, migrating to advanced OECD countries such as the USA and the UK. There is a high level of growth in some sectors such as computer programming and cybersecurity. There is also recruitment from advanced OECD countries such as Germany to the USA.

While companies may need such workers to fill certain jobs, the immigration policy of a country is also important as it sets out the legal rights for entry and residence for work purposes, e.g. through a visa or green card system. At a supranational level, freedom of movement of labour is one of the four economic freedoms of the 'acquis communautaire' of the EU. This encourages movement of skilled workers between European Union member states, e.g. French bankers working in the financial centre of London Docklands. Skilled migrants from developing countries may find it easier than non-skilled employees to work in a foreign country because of pro-immigrant policies of the host country. However, other factors may make it easier for these migrants to move, e.g. already established migrant communities and relocation support packages offered by companies. There might also be the chance of permanent residence for some workers as their working visa can be changed to a permanent one. Some countries such as Australia encourage skilled workers through a points system. In 2014–2015, there were nearly 200,000 places in Australia's migration programme and 68% of these places were taken by skilled workers. Some countries encourage foreign students with in-demand skills to stay on after they have finished their studies, rather than return to their home country or another recruiting country. For example, Japan's government has had a scheme whereby scholarships are offered to students. Government-sponsored job fairs, e.g. in Ireland, can guide workers more easily through the employment market and the visa required. This makes it more straightforward for migrants to cross borders.

However, there may be groups of migrants who are not classed as highly skilled but who may be able to have special consideration when moving across international borders. For example, the recent Syrian refugee crisis has prompted governments, e.g. within the EU, Turkey, Canada, to allow the legal residence of migrants who have had to flee the political situation there. The Australian government also has additional places within its humanitarian programme for the resettlement of refugees. However, it can be possible, as seen with the migrant crisis in Europe in 2015, for people to cross international borders without appropriate documentation.

This can put much strain on the migrants themselves as they seek to find a better life, as well as stresses for governments as they tackle the short- and long-term implications of migration. There may also be highly skilled migrants who do not find it easy to cross borders, e.g. those with criminal records or those who have wider family considerations.

It is easier in many cases for those with in-demand skills to move across borders than those with lower skill levels. However, some of these migrants may still have visa restrictions depending on their donor country's relationship with the host. In addition there are some lower-skilled workers who may find it easier to cross borders because of special agreements between countries, e.g. within the EU. Those with official refugee status may find it easier to move permanently to another country than those without.

ⓔ The student understands the question well and applies specific case study information from different countries and regions to explain and expand points. There is a clear focus on reasons why it is easier for highly skilled migrants to migrate, but also the fact that there may be other groups that may have easier access to foreign countries. The student attempts to draw a balanced conclusion. How the wealth or income of a person may affect their ability to migrate internationally could also be considered. **Level 4, 17 marks**

Student B

(a) (i)

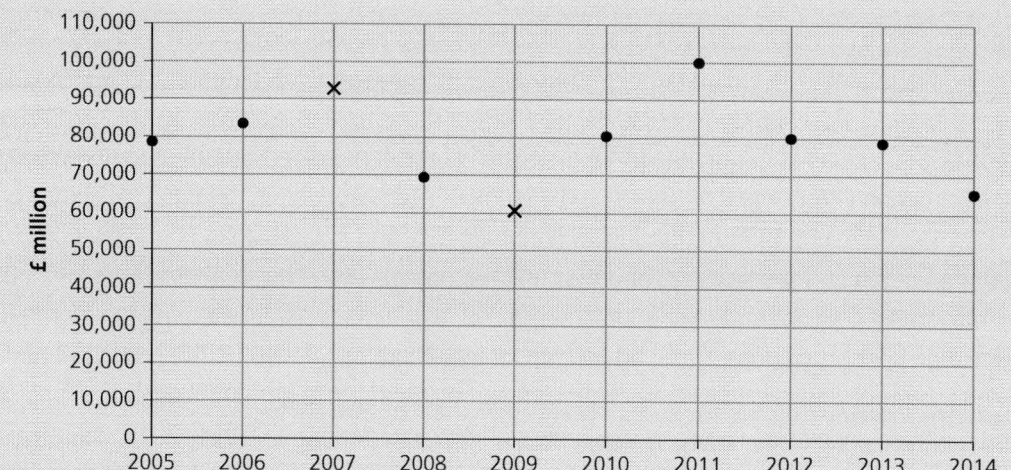

ⓔ Only one data point is correctly positioned. **1 mark**

(ii) $$\frac{78,000}{65,000} \times 100 = 120\%$$

ⓔ This is an incorrect calculation. **0 marks**

(b) The feeling of belonging to a particular country can be seen as part of your national identity. This can be complex in the UK as there are many different groups which identify with different countries within the UK rather than the UK as a whole. There are many rich people from Russia and the Middle East who own property in London. They can bring an international feel to the area and also bring money for the local economy. Although they live in London, they may not identify with being British. They may also keep themselves separate from other residents and this can change the sense of community of an area. Also some people have this as a second home for when they come to London on business. This means that when they are away, large numbers of properties are empty in residential areas. This is a change of tradition as in the UK people usually have their permanent residence near their place of work. This can have an impact on national identity.

(e) The student includes some good ideas linked to national identity and property but does not include specific examples related to business which would lift the answer into the Level 3 band.
Level 2, 4 marks

'complex in the UK' Valid point but not clearly linked to the question.

'own property in London' Example given of foreign ownership.

'keep themselves separate from other residents' Links to foreign ownership and sense of neighbourhood community.

'properties are empty' Links to foreign ownership and patterns of residence.

(c) Nationalism has an important influence on empire building. Nationalism is the concept of developing high levels of patriotism in a population and also promoting national values and culture. Nationalists often believe that their way of doing things is better than that of other peoples. Nationalism can lead to increased military power and support for military action abroad. This can also lead to conflicts with local people, e.g. colonial wars. Countries in the nineteenth century needed to have more raw materials and food supplies for a growing population because of the industrial revolution. If the resources could not be found at home, then leaders looked to expand their territory so that they could own or have priority over the resources. This was especially important as many European countries were competing for these resources. They also wanted control of expanding markets. Some countries also wanted more control over the trade routes in Southeast Asia, and to have more power than other nations in that region. This influence also led to colonies having to adopt the laws of the colonial power. In some cases this led to a loss of land ownership which had a negative impact on the local people. In the nineteenth century, many parts of the world were trying to become or did become independent from foreign rulers. For example, China fought Britain over disputes over trade of goods and China's sovereignty.

(e) **The student shows some knowledge and understanding but the answer is underdeveloped. China is used as an example but is not linked clearly to the question. The student could focus on a specific European power, e.g. France, and explain in more detail how an increase in nationalism encouraged the expansion of its territory to colonies for economic, cultural and political gain.**
Level 2, 5 marks

(d) Migration is the permanent or semi-permanent movement of people from one place to another. If a person moves between countries, they are an international migrant. Some people also want to move within a country. This is called internal migration. They may move from city to city to find work. People move for many reasons. They may want to find a job which pays more than their work at home, they may want to be with friends and family, or they can be forced to move if it is not safe for them to live in their home country. Many people feel that they are in danger because of war or because their views do not fit in with those of the government. That is why they move. Some people find it easy to move across a border. They have the right travel documents, e.g. a passport and the relevant visa needed to work in a country. These are legal workers. However, some people find it more difficult to move from one country to another. They may not get the official paperwork to move and have to pay traffickers large sums of money to get them where they want to go. This is dangerous and many people have died in the Mediterranean because of this. One example of where people were encouraged to move from one country to another was from Turkey to Germany. After the Second World War, the Germans had to rebuild many cities as they had been damaged in the war. There were not enough workers in West Germany to work in the construction industry as there was a lot of work to do and it needed to be done quickly. The problem was made worse by the Berlin Wall as workers from East Germany could no longer work in the West. In the 1960s, the West German government signed an agreement with many countries, including Turkey, to acquire workers for the construction industry. This was the 'gastarbeiter' scheme. Although these people had few qualifications, they had some skills and helped the Germans to improve their cities.

Today some Germans themselves migrate to other countries. Germany is a highly developed country with high levels of graduates. These graduates have developed high-level skills which are demanded by other countries. For example, the quality of German engineering is world famous. If a German engineer wants to live in the USA, he or she needs a green card.

In conclusion, it is much easier for people with higher skills to migrate to another country as they will find it easier to get a better job and won't rely on benefits from the government to survive.

e **The student shows some understanding of migration but ideas should be focused more clearly on the question (the international migration of highly skilled workers) rather than making general points. Internal migration is discussed, which is not directly linked to the question. The section on Turks in Germany shows some understanding but should have focused on highly skilled jobs rather than general construction. The student starts to show their knowledge of highly skilled workers (German engineers), but this section should be expanded to show more evidence of why some highly skilled workers can move more easily. The conclusion is brief.** **Level 3, 11 marks**

Synoptic investigation

Opportunities and challenges for Ethiopia

Question 1 mark scheme

4 marks (AO1 = 4 marks)

This question asks you to explain why some places are not as connected by globalisation as others. This lack of connectivity can be due to a range of factors including economic, social, physical, political or technological. You gain 1 mark for identifying a reason and a further 3 marks for expanding your answer.

For example:

Some countries are switched off economically from globalisation because they find it harder to access and benefit from world markets (1). Some countries are landlocked and so find it more difficult and expensive to ship their goods as they have to rely on good international relations with neighbouring coastal countries and may have increased transport costs (2). Some countries are cut off from other parts of the world because of the economic policies set by their governments (e.g. North Korea). This makes it more difficult for them to access free trade, which can increase their access to other markets (1). Less investment can come into the country, e.g. for transport networks, and this can make it harder for businesses to compete and increase their profits. Lack of investment in communications technology, e.g. smartphones and the internet, can also make it difficult for businesses to operate in a 24/7 global economy (1).

Other appropriate reasons will be accepted. You may use some of the data for Ethiopia from Section A of the resource booklet in your answer.

Hints and tips

Why do some countries find it difficult to connect to the global market? Think about access to free trade, level of technology, political isolation.

Question 1 example responses

Student A

Globalisation is the way in which different parts of the world are becoming more and more connected because of increased economic links though trade, as well as through sharing different cultural ideas and practices. Some countries do not have as much access through trade to the globalised economy as others and so stay mainly 'switched off'. These countries tend to be poorer nations that may not have coastal links and therefore find it harder to export goods (e.g. Chad). Some governments in these countries may have economic policies that are not focused on accessing free trade and which are less likely attract FDI. Also the economy of these countries may be structured around lower-value primary goods which may be subject to tariff barriers, making their products more expensive and less competitive globally. Local people who work in these sectors are likely to have lower wages and so global companies are less likely to invest as the local market has less spending power.

ⓔ This is a good answer that focuses on how economic factors can limit the access of poorer countries to the globalised world. Location, lack of transportation, government policy and market access are all discussed. An example of a relevant country is given. **4 marks**

'access through trade' Link to connectivity to world economy.

'may not have coastal links' Link to difficulties with transportation.

'economic policies' Link to economic policy.

'tariff barriers' Link to access to global markets.

Student B

Some locations find it hard to manage their natural resources. This might be because the government is not stable and there might be wars going on. Foreign companies do not want to invest when there is a chance of violence as they will not make a profit or keep their employees safe. The country finds it more difficult to export its products and so it cannot sell them as easily to the rest of the world as other more developed countries can. Some countries have bad environmental problems. For example, some countries in Africa have experienced drought and desertification.

ⓔ The student makes some connections between poorer countries and how they are at a disadvantage when accessing the global economy but does not always link their ideas clearly to the question. Focusing the answer around resources and the environment is good, but the student does not clearly show their understanding of how the issues mentioned affect access to the global economy. **2 marks**

'Foreign companies do not want to invest' Example of lack of foreign investment.

'difficult to export' Link to problems of accessing global markets.

'some countries in Africa' Not specific enough.

'drought and desertification' True, but not linked to the question.

216 © Hodder & Stoughton Ltd 2020   **Pearson Edexcel A-level Geography Exam Question Practice**

Mark schemes and example responses

Question 2 mark scheme

(a) 4 marks (AO3 = 4 marks)

You gain:

➤ 1 mark for completing the empty d and d^2 rows
➤ 1 mark for the sum of the d^2 column (Σ) = 32

Country	GNI per capita (PPP, US$)	Rank	KOF Index of Globalisation	Rank	d	d^2
Eritrea	1,130	10	27.13	10	0	0
Ethiopia	1,427	9	37.43	9	0	0
South Africa	12,100	6	64.82	3	3	9
UK	39,200	1	82.96	1	0	0
Algeria	13,054	4	49.36	8	4 (−4 also acceptable)	16
Egypt	10,512	7	56.33	6	1	1
Nigeria	5,341	8	54.05	7	1	1
Brazil	15,175	3	59.74	5	2	4
China	12,547	5	60.15	4	1	1
Turkey	18,667	2	69.02	2	0	0
					Σd^2	= 32

Sources of data: UN Development Report, 2015, and Axel Dreher, 2006

You gain 1 mark for correct working of the equation to find R:

$$R = 1 - \frac{6 \times 32}{10^3 - 10}$$

or

$$R = 1 - \frac{192}{990}$$

You gain 1 mark for the correct answer:

$$R = 0.81$$

or

$$R = 0.806$$

You gain 4 marks for the correct value of R alone. Errors can be carried forward at each step.

(b) 4 marks (AO3 = 4 marks)

Here you are asked to question the validity of the data you are using to make the calculation of R. You will gain 1 mark for identifying a reason why the data may not be reliable and 1 mark for expanding your idea, up to a maximum of 2 marks for each.

For example:

The data for GNI are average figures and will not reflect variations between different parts of a country, e.g. between the capital city and a remote region (1) or between different groups, e.g. very rich and very poor (1). Some areas/groups are more likely to be globalised/influenced by globalisation than others. The KOF Index consists of many different variables which are then put together in a formula (1). There may be inconsistencies between data sets for different countries, e.g. levels of accuracy and dates of published figures, and this will affect the validity of the data (1). Both data sets incorporate variables that are measured in US dollars. The US dollar, like all currencies, is subject to fluctuations in value (1). There are only ten countries and this is not enough to give a significant answer. There should be a minimum of 15 sets of data for a Spearman's rank correlation coefficient calculation to be valid (1).

You will not get credit for saying that you did not use the formula correctly or that your calculations were incorrect.

Hints and tips

Think about how the data are collected. Do the data include variables that may be more difficult to measure than others? Are there enough sets of data to make this type of calculation worthwhile?

Question 2 example responses

Student A

(a)

Country	GNI per capita (PPP, US$)	Rank	KOF Index of Globalisation	Rank	d	d²
Eritrea	1,130	10	27.13	10	0	0
Ethiopia	1,427	9	37.43	9	0	0
South Africa	12,100	6	64.82	3	3	9
UK	39,200	1	82.96	1	0	0
Algeria	13,054	4	49.36	8	–4	16
Egypt	10,512	7	56.33	6	1	1
Nigeria	5,341	8	54.05	7	1	1
Brazil	15,175	3	59.74	5	2	4
China	12,547	5	60.15	4	1	1
Turkey	18,667	2	69.02	2	0	0
					$\sum d^2$	= 32

$$R = 1 - \frac{192}{990} = 0.81$$

ⓔ **The student has completed all empty rows accurately and gives the correct value of R. Working is shown.** **4 marks**

(b) This example uses two secondary data sets (GNI per capita and the KOF Index of Globalisation) to calculate the Spearman's rank correlation coefficient. While both data sets are taken from the same year, there may still be <u>inconsistencies</u> with each data set which may affect their validity. Both data sets come from secondary data sources and so you are relying on another person, or organisation, to collect the data. It may be easier to collect data from some governments than others and the data itself may be gathered more accurately in some countries where <u>official record-keeping is more stringent</u>. Also, the KOF Index of Globalisation is made up of <u>many different components</u>: economic, social and political. There may be errors in data collection for some of these data sets in countries that are not so transparent and this will affect the reliability of the data. In addition, this is just one way of calculating globalisation. If another globalisation index that included <u>different variables</u> were used, then there may have been a different rank order.

ⓔ **This is a good answer that raises some specific points linked to problems with data validity. Ideas are well presented and show good understanding of the limitations of secondary data sets and multivariate indices.** **4 marks**

'inconsistencies' Problem with validity identified.

'official record-keeping is more stringent' Reason why data may have errors.

'many different components' Problems with consistency for multivariate indices.

'different variables' Problems of selecting data to measure globalisation.

Student B

(a)

Country	GNI per capita (PPP, US$)	Rank	KOF Index of Globalisation	Rank	d	d^2
Eritrea	1,130	10	27.13	10	0	0
Ethiopia	1,427	9	37.43	9	0	0
South Africa	12,100	6	64.82	3	3	9
UK	39,200	1	82.96	1	0	0
Algeria	13,054	4	49.36	8	4	16
Egypt	10,512	7	56.33	6	1	1
Nigeria	5,341	8	54.05	7	1	1
Brazil	15,175	3	59.74	5	2	4
China	12,547	5	60.15	4	1	1
Turkey	18,667	2	69.02	2	0	0
						$\sum d^2 = 24$

$R = 0.27$

(e) **The student has correctly completed the empty rows, but $\sum d^2$ is miscalculated so the value of R is incorrect. If the student had shown their working and used the correct method, an extra mark could be awarded even if the final value of R is incorrect.** **1 mark**

(b) The value may not be right because the person calculating R may have got their <u>figures muddled up</u> and so will have got the wrong answer. Some data may <u>not be measured correctly</u> or it may be difficult and expensive to collect. Some things are <u>easier to measure</u> than others, e.g. numbers of IKEA in the KOF Index is easier to count than gathering together the data needed to work out the GNI per capita. The number of <u>children enrolled in schools</u> is quite easy to collect as each school will have to fill in a list of who attends.

(e) **The student shows some ideas about why data used to calculate R are unreliable and gives some valid points about the difficulties in collecting accurate data for different variables. However, relevant examples are not always chosen. There is no need to include why the calculation itself may be wrong.** **2 marks**

'figures muddled up' Focus on the calculation itself and not the data used.

'not be measured correctly' Problems with data accuracy.

'easier to measure' Some data sets are easier to obtain accurately than others.

'children enrolled in schools' Point not linked to a data set in the question.

Question 3 mark scheme

8 marks (AO1 = 4 marks, AO3 = 4 marks)

This question asks you to look at selected data from four African countries and analyse the differences between their levels of development. Suggested ideas are outlined below, but you do not need to include all of these in your response. Other relevant points will be given credit.

AO1 Demonstrating your knowledge and understanding

➤ The level of development can be analysed using different variables such as economic, social and technological, but the validity of different data sets can be questioned as some may be unreliable.

➤ Higher levels of economic development can lead to investment in other aspects of development such as healthcare.

➤ Countries with young populations are more likely to have high birth rates and lower levels of economic development.

AO3 Using relevant skills to analyse data

➤ The data present a mixed view of levels of development for the selected countries.

➤ Algeria performs well in many of the indicators, e.g. it has the highest GNI per capita, the lowest dependency ratio, and life expectancy is relatively high. However, the internet-user rate is the lowest.

➤ Nigeria has the lowest GNI and performs poorly in terms of life expectancy and infant mortality.

➤ Egypt has the highest economic growth rate, indicating that there is potential for higher levels of development, although this might be focused on human development.

➤ South Africa, while having a high GNI, has a low life expectancy, suggesting that not all inhabitants benefit from the economy.

Answers to this question will be given a mark within a level band

Level 1 (1–3 marks) You show isolated knowledge and understanding of the differences in the level of development between the countries, with some inaccuracies. You make a limited attempt to analyse the data and make few geographical connections.

Level 2 (4–6 marks) You show some knowledge and understanding of the differences in the level of development between the countries, possibly with some inaccuracies. You make an attempt to analyse the data and make mostly relevant geographical connections.

Level 3 (7–8 marks) You show accurate and relevant knowledge and understanding of the differences in the level of development between the countries throughout. You analyse the data thoroughly and make relevant geographical connections.

Hints and tips

Look to see if there are countries that are more developed than others. Is it a clear decision? How can you use figures from the table to justify your points? It is unlikely that you will be able to comment on all of the data in the table in the time allowed, so be selective.

Question 3 example responses

Student A

The four countries shown have different levels of development. It could be argued that out of these countries, Algeria is the most developed. It has the highest GNI per capita so the country will have more to spend on healthcare and education. This could be shown by the fact that Algeria has the highest life expectancy of this group (74) and one of the lowest infant mortality rates (21 per 1,000). Out of the group, Algeria has the lowest dependency ratio for young people. This indicates that while there is a potential workforce for the country, the burden of providing for the needs of young people, e.g. schools, is not as great as in the other countries. However, only 18% the population are internet users, which might suggest that the country is not investing in communications technology to meet demand.

Egypt could be seen as at roughly the same level of development as Algeria. Although its GNI is lower, its economic growth rate is slightly higher, indicating some potential for future development. Life expectancy is high for the group and its infant mortality rate is the lowest (18 per 1,000).

South Africa also has a high GNI but a much lower life expectancy and a higher infant mortality rate. This might imply that the wealth created in South Africa does not have a direct effect on the lives of many of its people. Some groups here may be particularly disadvantaged.

Nigeria is the least developed within this group. Economically, it has the lowest GNI and the lowest annual growth (2.7%). Healthcare provision is poor as, strikingly, the infant mortality rate is extremely high and life expectancy is low. However, the number of internet users is high, suggesting that the government may have prioritised this in its planning. While its high population on its own is not an indicator of development, it does show that if more money is invested in healthcare, the potential for social development is huge.

ⓔ The student analyses the data effectively and tries to assess which countries are more developed than others. The student shows good understanding and makes connections between the data and their own knowledge. Many examples from each country are given to justify points. A suggestion that level of development goes beyond just the economic indicators is also made. **Level 3, 8 marks**

Student B

The four countries are all part of the African continent, with Algeria bordering the Mediterranean and South Africa being the furthest south. Different indicators have been used to see whether the countries are developed or not. From the data we can see that Algeria is the most developed country and Nigeria is the least developed country.

Algeria is the richest country as it has the highest GNI. This means that there are wealthy people in Algeria. They can afford to live in better houses and eat more healthily. Because people have a better standard of living, life expectancy is high. Also fewer babies die than in other countries. This means that its healthcare system is quite good. Either the government can afford to provide better health facilities or the people are able to pay for them privately.

Nigeria is the poorest country but it has the highest population. This has serious consequences for development as many people will be affected if the quality of life in the country is poor. Nigeria has the lowest GNI and this is half of the other countries, so it is significantly poorer. The quality of life for people is not very good. Life expectancy is only 52. Also, many babies die early as the infant mortality rate is way higher than the rest. It may be difficult for Nigerians to afford the healthcare that they need. However, there is a relatively good level of internet access, although this might not be the same throughout the country.

ⓔ **The student explains their ideas well in general for Algeria and Nigeria. Connections between the data and the student's own knowledge are made, and logical ideas put forward. Relevant data are used to justify points. However, the student states that Algeria has the best score for each indicator, but this is not the case. There is no comment on the data for Egypt or South Africa, which would have shown more thorough analysis.** **Level 2, 5 marks**

Question 4 mark scheme

8 marks (AO1 = 4 marks, AO3 = 4 marks)

This question asks you to analyse the relationship between the HDI and energy use per capita for selected countries. In general the higher the energy use per capita, the higher the HDI, for economic, social, environmental and political reasons. You should show some understanding of the graph as well as suggesting reasons for the relationship shown. Suggested ideas are outlined below, but you do not need to include all of these in your response. Other relevant points will be given credit.

AO1 Demonstrating your knowledge and understanding

➤ The Human Development Index combines variables from key aspects of human wellbeing. It shows attainments in life expectancy, education and standard of living.

➤ Access to energy is important for people's overall quality of life. Having a good energy supply can help to improve living standards, industrial capacity and transportation.

➤ In general, the higher the HDI, the more energy is used per capita.

➤ The amount of energy used depends on its availability. This is linked to how much energy is produced within the country itself, whether it needs to import energy, the type of energy it uses (renewable or non-renewable), energy costs and government policy.

AO3 Using relevant skills to analyse data

➤ There is a positive correlation between HDI and energy use per capita. In general, as HDI increases, so does energy use per capita.

➤ Those countries with a low HDI tend to have lower energy use. These countries may find it expensive to import oil-based products and so rely more on renewable energy, e.g. Ethiopia.

➤ Emerging countries such as Mexico and Chile have higher energy use per capita.

Synoptic investigation

➤ There are anomalies in the lower HDI group, e.g. South Africa has a lower HDI than other emerging countries but a higher energy use, possibly because of inequalities between different regions and different people.

➤ Advanced economies have a higher HDI and higher energy use, e.g. the USA, where there is an advanced standard of living and high levels of car usage. Iceland has a significantly larger energy use per capita because of its abundant sources of renewable energy.

➤ Some data may be unreliable as they may have been collected at different times.

➤ There are only 11 countries shown in Figure 9. A much larger data set would be needed to make more qualified statements about the relationship.

Answers to this question will be given a mark within a level band

Level 1 (1–3 marks) You show isolated knowledge and understanding of the relationship between HDI and energy use per capita, with some inaccuracies. You make a limited attempt to analyse the data and make few geographical connections.

Level 2 (4–6 marks) You show some knowledge and understanding of the relationship between HDI and energy use per capita, possibly with some inaccuracies. You make an attempt to analyse the data and make mostly relevant geographical connections.

Level 3 (7–8 marks) You show accurate and relevant knowledge and understanding of the relationship between HDI and energy use per capita throughout. You analyse the data thoroughly and make relevant geographical connections.

Hints and tips

What is the relationship shown? Can you suggest reasons for the position of different countries on the graph? Why might the data be unreliable?

Question 4 example responses

Student A

There is a clear relationship shown by the data in Figure 9. In general, the higher the HDI for a country, the higher its energy use per capita. There is a positive correlation between the two variables, although only 11 countries have been included. The data from more countries would be needed for the relationship to be tested statistically.

From the data shown, most countries with HDIs of under 0.7 have energy uses of 1,000 kg of oil equivalent or less. This is because access to reliable and affordable energy is highly important when developing services and infrastructure to improve people's quality of life. However, South Africa's energy use is much higher (around 2,700 kg of oil equivalent) while its HDI is under 0.7. This may be because the South African economy uses a significant amount of energy while the benefits of this do not trickle down to all sectors of society. They may also produce energy themselves and so do not have to spend money on costly oil imports.

There are also some differences between those countries with HDIs above 0.8. Chile has a relatively high HDI but uses much less energy per capita than more advanced nations, e.g. the USA. This may be because economies such as the USA require higher amounts of energy for their cars and to maintain their high quality of life. Finally, Iceland uses a significantly higher amount of energy than the other advanced nations shown (around 18,000 kg of oil equivalent). This may be because Iceland has vast quantities of renewable energy at its disposal. It also may have to use large quantities to maintain high standards of living in such a cold environment.

ⓔ The student understands the data and explains the relationship well. The student attempts to group the countries to help put forward points and point out anomalies. Connections are made between the data and the wider issue of the need for energy to maintain standards of living and to fuel the economy. Valid reasons for some countries' positions on the graph are given. Good use is made of examples of numerical data to justify points. The validity of the data presented is questioned. A concluding sentence to summarise the ideas would improve the answer. **Level 3, 7 marks**

Student B

The graph shows information for ten different countries. These countries are from all around the world and are both rich and poor. The graph compares the energy use per capita with the HDI value for each country. There are some interesting points when you look at the graph. For example, Iceland is a rich country with a high standard of living. This is shown by the high HDI of around 0.9. It also uses way more energy than all of the other countries. This might be because there are a lot of volcanoes in Iceland and so there is a lot of geothermal energy. It can use this energy to make electricity and to heat water. It is also cheap and renewable and is there all the time so they can use as much of it as they want. On the other hand, Ethiopia is a poor country with limited resources. It has a low HDI because many people do not have access to the things they need to have a good life, e.g. clean water and schools. Because people don't have much, e.g. no car, they do not need to use so much energy. However, some countries that have a slightly higher HDI still use under 2,000 kg. This may be because they do not need as much energy as the richer countries but their industry is starting to develop and so they need more. As the HDI goes up, so does the energy use per person.

ⓔ The student understands the main relationship shown in the graph and generally makes some valid connections to their own knowledge about energy use. However, too much time is spent on the Iceland example. A better approach would be to group countries with similar characteristics together, which would help broaden the answer. An attempt to use examples of the data to help explain points is made and a summary is given. The number of countries cited is incorrect. **Level 2, 5 marks**

Question 5 mark scheme

18 marks (AO1 = 3 marks, AO2 = 9 marks, AO3 = 6 marks)

This question asks you to evaluate different aspects of Ethiopia's energy sector, including energy production and consumption. Suggested ideas are outlined below but you do not need to include all of these in your response. Other relevant points will be given credit.

AO1 Demonstrating your knowledge and understanding
➤ The energy sector plays an important role in the development of a country as many improvements depend on an affordable and reliable energy supply.
➤ Developing countries are encouraged to find sustainable ways to meet growing demand, e.g. renewable energy.

AO2 Applying your knowledge and understanding
➤ Ethiopia's dependency on renewables allows it to produce greener energy with low carbon emissions.
➤ Ethiopia does not import a large proportion of its energy, which makes it less susceptible to fluctuations on the world energy market.

- ➤ Many citizens still do not have access to electricity so do not use it for daily living. This may affect their standard of living and put pressure on other resources, e.g. fuelwood.
- ➤ There are considerable plans for substantial new energy development based around renewable energy. There are benefits and costs in investing in large dam schemes.
- ➤ Increasing foreign interest in the energy sector may bring opportunities and challenges.

AO3 Using a variety of different skills to analyse data

- ➤ Figure 10 suggests that renewable energy and in particular HEP plays an important part in Ethiopia's energy production today.
- ➤ Figure 11 suggests that Ethiopia is looking to expand its energy production and to improve access to electricity for its citizens, helping to improve quality of life.
- ➤ Figure 12 suggests that Ethiopia is focusing on developing renewable energy in the future. The energy projects produce different amounts of electricity and some are more costly than others. Some projects have been connected to foreign firms for financing and expertise.

Answers to this question will be given a mark within a level band

Level 1 (1–6 marks) You show limited knowledge and understanding of Ethiopia's energy sector, with some inaccuracies. You apply your knowledge with few connections and with limited support. Your conclusion, if any, is unbalanced. You make judgements that are not supported by relevant evidence. Your interpretation of the evidence does not show a clear connection with ideas from the geographical topics studied.

Level 2 (7–12 marks) You show knowledge and understanding of Ethiopia's energy sector, possibly with a few inaccuracies. You apply your knowledge with some connections and with some support. You draw a conclusion but your arguments may be unbalanced. You make judgements that are sometimes supported by relevant evidence. Your interpretation of the evidence does show some connections with ideas from the geographical topics studied.

Level 3 (13–18 marks) You show accurate and thorough knowledge and understanding of Ethiopia's energy sector. You apply your knowledge with many connections and with much support. You draw a logical, substantiated conclusion and your arguments are balanced. You make judgements that are always supported by relevant evidence. You critically interpret the evidence and show clear connections with ideas from the geographical topics studied.

Hints and tips

Try to work out the strengths and weaknesses of the energy sector. What types of energy are produced? Does the population have access to electricity? Are there any future issues which may arise from Ethiopia's energy plan?

Question 5 example responses

Student A

Access to reliable and affordable energy sources is highly important for a country that wants to improve its level of overall development. Energy is important for the economy and society of a country. Economically, countries that can generate their own energy supply, or indeed produce enough to export, are in a much better position than those that cannot. This is because they are more likely to be energy secure and not susceptible to changes in the world market. If the energy produced is green energy, i.e. that which has low levels of carbon emissions, then this benefits the country's environment and can be beneficial for the health of its citizens. However, there remain challenges for countries that produce renewable energy, such as reliability concerns as well as the cost-effectiveness of large dam projects.

Ethiopia is a low-income country that already produces a large amount of the energy it needs from renewable resources, in particular HEP. It is clear that much more energy will be needed in the future as only 26% of Ethiopians have access to electricity. Once more people are able to gain access to the grid, even for basic needs like cooking and lighting, then demand will rise. Although Ethiopia only has to import 5% of the energy that it uses at the moment, more energy-generation schemes will have to be created in the future if increased demand is to be met without relying on expensive imports. The government has already started to improve connectivity in rural areas through its UEAP scheme and energy is a focus of its GTP. Figure 12 highlights a range of current and future investments in renewable energy schemes. The most significant of these are the large multi-billion-dollar dam schemes, e.g. GERD, which aim to produce the increasing levels of energy that Ethiopia demands. Foreign partnerships, e.g. with Iceland, help Ethiopia to develop geothermal energy. The Great Rift Valley is located in Ethiopia and there is much potential for geothermal energy exploitation here.

However, there are some challenges which remain. Large dam schemes such as GERD and GIBE are expensive and Ethiopia may have to borrow money to afford them. Some may argue that part of this money should instead go on projects which directly improve the education and health of the population. Also, HEP relies on regular precipitation to ensure that rivers are full enough to power the hydroelectric turbines. Fluctuations in climate may influence the MW capacity of the HEP projects. Also, the GERD dam may alter the flow of the Blue Nile and lead to changes in river discharge further downstream in other countries. If the supply of water changes, then there may be tensions between Ethiopia and countries such as Egypt and Sudan that use water from the Nile. Large energy projects often require the expertise and financial assistance of other countries. This may affect the level of control that the government has on the energy sector. Government investment in training skilled professionals in this sector will be important for future growth. Finally, the transport sector is likely to grow and at the moment this relies heavily on oil, which Ethiopia imports. It could become increasingly vulnerable to changes in oil prices, unless more people in Ethiopia use electric transport in future.

Ethiopia's energy sector, like that of all countries, has strengths and weaknesses. However, the outlook is positive as the government is investing in renewable technology to meet the inevitable growth in demand over the next decade. There is much potential and Ethiopia is in a good position to capitalise from the energy resources that it has. However, careful planning to limit the negative impacts of these schemes, e.g. political tensions and environmental concerns, will be needed for this potential to be fulfilled.

ⓔ **The student uses a range of evidence from Section C to develop a good evaluation of Ethiopia's energy sector. They clearly understand the importance of renewables to Ethiopia, both now and in the future. Strengths and weaknesses of the sector as a whole, such as the population's access to the grid, and also more specifically of the renewables sector, are outlined. The arguments are balanced and a judgement is made in the conclusion. How Ethiopia is 'leading the way' with some types of renewable development, e.g. the geothermal and waste-to-energy schemes, could be mentioned.** **Level 3, 16 marks**

Student B

Ethiopia is a poor country and its inhabitants do not use much energy. Most people cook with wood and they do not use electricity in their daily lives. This is probably because most people find it difficult to get electricity and live in rural areas which are far away from a connection. So the government decided to invest in a lot of energy projects to bring more electricity to the people. Some of these are shown in Figure 12. Renewable energy is a great way to produce energy. It uses naturally available resources which do not run out and it does not harm the environment. For a country like Ethiopia which has a large population, this is ideal. Ethiopia has invested in HEP projects because it already produces a lot of energy this way and so has a wealth of experience. It is also lucky enough to have the River Nile, one of the longest rivers in the world, flowing through it. This gives Ethiopia a ready-made source of water which can be used to produce lots of MWs, exactly what Ethiopia needs. Ethiopia is also investing in wind turbines. Electricity can be generated by the flow of wind through a wind turbine and can then be sent on to the grid. It does not produce carbon dioxide emissions, which cuts down on greenhouse gas emissions. This helps to reduce global warming.

Geothermal energy is also limitless once you have built the plant. This happens when steam from the ground is used to drive a turbine which then produces electricity. By constructing the Corbetti plant, the government can use the energy found naturally under the earth to generate electricity and provide hot water. Finally, the Reppie scheme is an excellent approach to generating sustainable energy. The project is so good it has been backed by Barack Obama. The idea is that waste from people in cities can be converted into energy that then can be used for other things. It will be situated in the capital city of Addis Ababa. This is a good location for this scheme as lots of people live here and produce lots of rubbish. If the scheme works well here, it could be a solution for both rubbish removal and energy production in other African cities. However, the plant only produces a relatively small amount of power and so would only contribute a little to the overall energy capacity.

The Ethiopian government has also had help from other countries to build its projects. The Chinese have helped it with the GIBE III dam. This may be because the Chinese have a lot of experience with building large dams such as the Three Gorges Dam. In addition, Iceland has a large amount of geothermal energy and it can use its expertise to develop plants in Ethiopia using the latest technology.

Ethiopia has a healthy energy sector. It produces most of what it needs through renewable energy. This is good for the environment and good for the people of Ethiopia.

ⓔ **The student demonstrates understanding of key concepts and attempts to interpret some of the sources. Some strengths and weaknesses are given but points are not always linked together well. The focus of the answer is drawn from Figure 12; consideration of the rest of Section C would help to balance it. Strengths of renewables are discussed more than weaknesses; this affects the structure of the response and the student's ability to convey their arguments. The conclusion is a little brief. The answer would be improved by drawing together the main arguments to make a clear judgement at the end.** **Level 2, 10 marks**

Question 6 mark scheme

24 marks (AO1 = 4 marks, AO2 = 12 marks, AO3 = 8 marks)

This question asks you to consider whether you think Ethiopia is in a strong position to become one of Africa's leading economies. You should use information from all sections of the resource booklet. Suggested ideas are outlined below, but you do not need to include all of these in your response. Other relevant points will be given credit.

AO1 Demonstrating your knowledge and understanding

➤ Ethiopia has both opportunities and challenges that will affect its potential to become one of Africa's leading economies.

➤ There are economic opportunities, such as the growth of the energy and tourism sector, as well as economic challenges, e.g. low average incomes and poor infrastructure.

➤ There are political opportunities, such as the role of Addis Ababa in regional politics, and social challenges, such as a high youthful dependency ratio and food security issues.

➤ There are environmental opportunities, such as the potential for green energy, but also environmental challenges, e.g. the threat of drought.

AO2 Applying your knowledge and understanding

Some possible arguments for Ethiopia being in a strong position:

➤ Good economic growth with increasing levels of FDI.

➤ Potentially large young and dynamic workforce.

➤ Improvements in transport and communication technology.

➤ Government-led planning and investment, particularly in renewable energy and tourism.

Some possible arguments for Ethiopia being in a weak position:

➤ Landlocked country.

➤ Still large sectors of the population with low living standards and limited access to healthcare.

➤ High drop-out rate after primary school, which could impact on literacy levels.

➤ Susceptibility to droughts causing significant challenges for the population.

➤ Balance of trade based on more expensive imports, e.g. oil.

➤ Still lags behind other African populations in many economic and social indicators.

➤ Possible tensions between different groups.

AO3 Using a variety of different skills to analyse data

➤ Figure 1 and Section D: Ethiopia is a landlocked country that is becoming increasingly more globally connected through international air and rail links.

➤ Figure 3: Ethiopia has had an unstable recent past but there are signs of structured recovery.

➤ Figure 4: Ethiopia performs weakly in many social indicators. The population is large compared with some other African countries (Figure 8). Issues with education and health limit population potential.

➤ Section C: Energy mix is mainly based around the renewables sector. Low per capita energy usage and limited connectivity for some households. Foreign partnerships are often involved.

➤ Section D: Coffee production by cooperatives, which also invest in the community. Growth of the tourist sector. Severe impacts of recent droughts affecting many parts of the country. Many groups are vulnerable to changes in climatic patterns.

Answers to this question will be given a mark within a level band

Level 1 (1–6 marks) You show limited knowledge and understanding of Ethiopia's position and whether it can become one of Africa's leading economies, with some inaccuracies. You apply your knowledge with few connections and with limited support. Your conclusion, if any, is unbalanced. You make judgements that are not supported by relevant evidence. Your interpretation of the evidence does not show a clear connection with ideas from the geographical topics studied.

Level 2 (7–12 marks) You show knowledge and understanding of Ethiopia's position and whether it can become one of Africa's leading economies, possibly with a few inaccuracies. You apply your knowledge with some connections and with some support. You draw a conclusion but your arguments may be unbalanced. You make judgements that are sometimes supported by relevant evidence. Your interpretation of the evidence shows some connections with ideas from the geographical topics studied.

Level 3 (13–18 marks) You show mostly accurate and relevant knowledge and understanding of Ethiopia's position and start to evaluate whether it can become one of Africa's leading economies. You apply your knowledge with some connections and with some support. You draw a conclusion that is supported by an argument which is sometimes unbalanced. You make mostly valid judgements that are mostly supported by relevant evidence. You critically interpret the evidence and show some clear connections with ideas from the geographical topics studied.

Level 4 (19–24 marks) You show accurate and thorough knowledge and understanding of Ethiopia's position and evaluate whether it can become one of Africa's leading economies. You apply your knowledge with many connections and with much support. You draw a logical, substantiated conclusion and your arguments are balanced. You make judgements that are always supported by relevant evidence. You critically interpret the evidence and show clear connections with ideas from the geographical topics studied.

Hints and tips

Use all the resources. Work out what are the strengths and weaknesses of Ethiopia's current position. Do you think it has the potential to become a leading African power?

Question 6 example responses

Student A

Ethiopia is one of the world's poorest countries with an HDI of 0.442. Since the Second World War, its development has been negatively affected by the role of different political groups, especially the Derg government in the 1970s and 1980s, as well as the legacy of the severe drought in the 1980s. Although it faces severe economic, social and environmental challenges today, recent developments have led some commentators to suggest that Ethiopia is one of Africa's 'Big Five' countries and so would be in a position to develop to become a leading economy within the continent. A more detailed assessment of the strengths and weaknesses of Ethiopia's current situation is required to determine its future prospects.

An economy based on primary exports (e.g. coffee) and expensive imports (e.g. petrol and machinery) makes the country more vulnerable to changes in world markets. However, Ethiopia is focused on trying to produce quality coffee, with a fair income for its farmers. Organisations like the OCFCU enable farmers to gain finance to help them grow their products organically and to promote them abroad using the Rainforest Alliance scheme. Farmers get a decent share of the profits and also the OCFCU can invest in educational projects such as new primary schools and early years care. This will help improve access to education, something that will need to be targeted if Ethiopia is to improve its economy and status. Ethiopia also is trying to diversify its economy to increase its income potential and make it less vulnerable to economic shocks in one particular sector. Ethiopia is developing its transportation networks to encourage the flow of goods, and the tourism sector is growing. A new train link to Djibouti port will give Ethiopia the ability to export larger quantities of bulky goods, and for foreign companies to access the market more easily. Government-owned Ethiopian Airlines has also grown in scale and is becoming a major African player in the market.

A reliable modern airline is important not just for trade but also for the impact on a country's image, something that will be important if Ethiopia is to add to its status. The number of tourists who come to Ethiopia is increasing, many to see the beautiful landscape and the nine UNESCO sites. Continued development of this will add more foreign exchange to the economy and also raise Ethiopia's profile abroad in a positive way.

In order to achieve sustainable economic growth, Ethiopia has invested in more renewable energy projects with the goal of being able to export energy. The country already has a very green energy mix, with HEP accounting for 88% of power. Ethiopia has an abundance of natural resources from which to gain energy, including geothermal power. Projects like the Corbetti plant will help produce sustainable energy. However, there has been a large amount of investment in large-scale dam schemes. While this has the benefit of producing large quantities of HEP, problems such as unsustainable costs, unreliable rains and displacement of local people are present. On an international level, the GERD and GIBE III schemes affect rivers which then flow into other countries. Changes in downstream flows could lead to international disputes that may be difficult to solve.

Despite these encouraging signs of growth, Ethiopia will have to overcome some major challenges if it is to become a leading power in Africa. Its relatively large population (102.3 million) is very young and many children leave the education system early without the necessary skills that will be needed to benefit the economy. Much of Ethiopia's population is poor and with little access to basic services. The low quality of life is highlighted by low life expectancy and high infant mortality rates. However, healthcare is improving as the death rate is now around 8 per 1,000. Certain groups may be particularly disadvantaged. For example, rural populations are cut off from the electricity grid and ethnic groups have held demonstrations over land rights. At the moment, there are extremely low levels of motor vehicle ownership. If this rises because of increased wealth, then Ethiopia will have to deal quickly with issues such as emissions and congestion on major routes. Food insecurity and water shortages caused by droughts, and problems with logistics and supply of these basic resources, also add to the country's vulnerability.

However, other evidence would also be useful when considering whether Ethiopia is in a position for growth, such as information on the amount earned and the number employed in each economic sector. Also, more information about the different ethnic groups could provide more insight into potential inequalities which might slow down growth.

Overall the evidence suggests that Ethiopia has the potential to become a leading economy, mainly because of its potential in the renewable energy sector. Not only will this produce the energy needed for growth but it could also earn money from exports. However, when compared with other members of the 'Big Five', Ethiopia has a long way to go to match or even surpass the size of their economies (based on the GNI figures) and has significant social challenges to overcome. How the government manages these challenges will play a key role in the country's future success.

ⓔ **A good level of knowledge and understanding is shown about the possibility of Ethiopia becoming a leading economy. The data are critically analysed. A balanced argument is offered that clearly conveys Ethiopia's strengths and weaknesses. Relevant evidence, selected from across the resources, is used. The data are challenged by suggesting that other factors should be considered in assessing Ethiopia's potential. There is a balanced conclusion and a judgement linked to the arguments. Additional marks could be gained by noting Ethiopia's political status in hosting organisations (e.g. African Union), and the money brought in by foreign diplomats and decision makers.** **Level 4, 20 marks**

Student B

Ethiopia is a poor country in Africa that is trying hard to develop. Like many countries, it has advantages which it can use to help it become one of Africa's leading economies, as well as major issues it has to overcome to help it get there.

Ethiopia is situated in northeast Africa but has no coastline of its own. Its physical landscape is divided into three main zones, including highland and lowland areas. These areas receive different amounts of rainfall, with the western side getting much more rain than the eastern side. However, sometimes this rain is unreliable. Indeed in 1984 there was a catastrophic drought, which caused the crops to fail. Many people died of hunger and the world raised money through the Live Aid concert. Today, there are still problems with drought as over 10 million people need food aid. This is about 10% of the population. People are going hungry and it is difficult for some mothers to breastfeed their babies. Having a drought means that the water supply dries up and this can lead to the spread of disease as people cannot look after themselves properly. Water has had to be driven into some regions using trucks. This help comes from outside sources as Ethiopia cannot provide all of this assistance itself. This makes it less likely to be a leader as it does not have enough money to deal with these emergencies.

The lack of rainfall also makes it difficult for the energy industry. Ethiopia has ambitious plans to develop its economy and will need to have more power to do this. If it produces a lot of energy, then it will be able to export this and this will help it to earn money. In addition, if it has energy security, it will be in a better political and economic situation compared with those African countries that do not. Most of its energy comes from HEP at the moment and it has plans to develop this sector. However, if the rainfall becomes unreliable, then there might not be as much energy produced from these schemes. HEP is an expensive investment and there might be financial issues if there is not enough energy to sell at a profit.

If Ethiopia is to become a leading African economy then it will have to improve the quality of life for its people. At the moment, its GNI per capita is much lower than that of the other 'Big Five' countries. This reflects the wealth of the country overall, which has an impact on how much money the government has to run the health and education schemes needed to improve people's standard of living. The literacy and secondary school attendance rates are low and this will have an impact on the future economy. There are large numbers of young people in the population at the moment. If these people are not educated, they may find it difficult to get a job. It also might slow down the economy as Ethiopia will not have the skilled workforce it needs and may find it difficult to attract people from abroad because wages are low. It has few internet users at the moment but it is hoped that this will increase. However, Figure 19 shows that money can be sent by mobile phones. This might be useful as the economy starts to grow as not everywhere might have a bank and it might not be safe to keep money at home. It will mean people can manage their money more effectively. This technology will also be useful if Ethiopian companies want to deal with foreign companies and can do business more easily, particularly in rural areas. However, with poor access to electricity for many rural people, charging up phones may be difficult at times.

However, Ethiopia's tourist industry is booming. There has been a significant growth in the number of tourists and the country has invested in its aircraft and airports. This will bring much-needed foreign currency into Ethiopia and provide employment opportunities for people, particularly outside the big cities in roles such as drivers, guides, hotel and restaurant workers. The growth of this sector may help Ethiopia's plan to become a more significant African economy.

All in all, I do not think that Ethiopia will become one of Africa's leading economies, in the near future at any rate. It has major economic and social hurdles to overcome before many people's standard of living improves, particularly in health and schooling. People in the countryside have limited access to electricity, which will have to change if Ethiopia is to grow economically. It is starting to develop new sectors of the economy but these will have to be successful for a long time to earn enough money for Ethiopia to be classed as one of Africa's leading economies.

(e) Some of the resources are well interpreted, but the points made and evidence selected do not always link clearly to the question. There is occasional over-reliance on the resources. Some geographical understanding is shown and there is an attempt to establish strengths and weaknesses. However, the focus is mostly on weaknesses, giving an unbalanced answer. Knowledge is sometimes applied but occasionally lacks clarification relating to the question. A judgement and connections with other parts of the course are attempted. A discussion of more ways for Ethiopia to increase its economy and standing, e.g. positive aspects of energy projects, home of AU, would improve the answer. **Level 3, 14 marks**

Opportunities and challenges for the Arctic region

Question 1 mark scheme

4 marks (AO1 = 4 marks)

This question asks you to explain why the cryosphere is an important water store. A large proportion of the planet's fresh water is held within a range of locations in cold environments. You gain 1 mark for identifying a reason and a further 3 marks for expanding your answer.

For example:

The cryosphere consists of water found on the Earth's surface as ice. Only 3% of the planet's water is fresh and 79% of this is locked up in ice sheets, ice caps and glaciers (1). Major river systems are fed by melted ice from the cryosphere (1) and so water that is stored in ice forms a large proportion of our potential fresh water supply (1). Ice in the cryosphere, for example in the Greenland ice sheet, may have built up over many thousands of years and so once used would not be able to be replaced in a realistic timeframe (1). A decline in the volume of ice in the cryosphere would have an impact on the availability of fresh water and could lead to severe reductions in quality of life for people and also affect other sectors such as farming and some manufacturing. A reduced supply of water could also lead to political and/or military conflicts.

Other appropriate reasons will be accepted. You may use some information from Section A of the resource booklet in your answer.

Hints and tips
How significant is the cryosphere in terms of the planet's water stores? How does the cryosphere link to our fresh water supply?

Question 1 example responses

Student A

The cryosphere is a significant part of our planet's water storage system. Fresh water is held within the ice found in polar and alpine regions. For example, <u>nearly 80%</u> of the world's fresh water is frozen as ice within ice caps, such as in Iceland, and glaciers. As fresh water only accounts for <u>3% of all the Earth's water</u>, the contribution that ice makes to this amount is very important. When ice melts, the water that is released can be used for many types of activities. One major use is the consistent supply of fresh water to the <u>world's population</u>. Ice from many of the world's major glaciers feeds river systems which have large dependent populations. <u>This is essential</u> for safe, clean drinking water as well as for general health and sanitation. Any decrease in the amount of ice in the cryosphere would lead to a reduction of water availability, especially as the ice has built up over 10,000 years in some cases and it would be difficult to replenish.

e **This is a good answer that focuses on why the cryosphere is a key water source. Its importance is highlighted both within a global context and with other sources of fresh water. The student expands their answer by linking it to water use and quality. A relevant named example is given.**

4 marks

'nearly 80%' Point made that a significant proportion of fresh water is in the cryosphere.

'3% of all the Earth's water' Puts into a global context.

'world's population' Links to needs of population.

'This is essential' Expands importance for quality of life.

Student B

The cryosphere is <u>water that is kept in or on the ground as ice</u>. It can be found in many different locations within cold environments. For example, it can be found in the soil where it is known as permafrost. In addition, large areas of ice are found in <u>ice sheets in Greenland and ice caps in Iceland</u>. Ice is important as it is made up of frozen water. When the temperature increases, <u>ice melts and then flows into rivers and streams</u> where it can be used for drinking water. This is important for people to live and work.

e **The student shows understanding of what the cryosphere is and gives some locations but has not really put across its importance as a water store, which is demanded by the question. They could explain its importance in terms of scale and/or of freshwater storage. By using the resource booklet Section A, the student could highlight that a large proportion of fresh water is stored within the cryosphere.**

2 marks

'water that is kept in or on the ground as ice' General understanding of cryosphere.

'ice sheets in Greenland and ice caps in Iceland' Examples of locations given.

'ice melts and then flows into rivers and streams' Importance of cryosphere as an important fresh water source for the population.

Question 2 mark scheme

(a) **(i)** 2 marks (AO3 = 2 marks)

You gain 1 mark for showing your working:

$$\frac{14.5 - 16.5}{16.5} \times 100$$

You gain 1 mark for the correct answer:

$$= -12.12\%$$

(ii) 2 marks (AO3 = 2 marks)

You gain 1 mark for showing your working:

$$3.7 - 1.8$$

You gain 1 mark for the correct answer:

$$= 1.9 \text{ m}$$

For 1958–1976, a range of 3.6 to 3.8 will be accepted. For 2003–2007, a range of 1.7 to 1.9 will be accepted. A correct answer based on your own figures will be accepted if they are within the acceptable range.

(b) 4 marks (AO3 = 4 marks)

This question asks why the data used to calculate the values of sea ice extent and thickness may be unreliable. You gain 1 mark for identifying a reason why the data may not be reliable and 1 mark for expanding your idea (up to a maximum of 2 × 2 marks).

For example:

There may be differences in the accuracy of the data given because of changes in technology. Some of the data in both graphs were collected before satellites were used. Data collected by submarine or aircraft may not be as accurate as they have to cover sample areas rather than have a view of the whole Arctic Sea (1). It is difficult to monitor the exact state of the ice at any given time because the environment changes constantly. Samples of data may be used in models to give an idea of changes over time (1). Different groups may collect data with different methodologies, making direct comparisons difficult (1). Satellite imagery may be analysed in different ways, opening up the possibility of a range of results from the same data set (1).

You will not get credit for saying that your calculations were incorrect.

Hints and tips

Think about how the data are collected. How does the frequency and time of data collection have an impact on the results? Who might collect the data and how might this affect results?

Question 2 example responses

Student A

(a) (i) $\dfrac{14.5 - 16.5}{16.5} \times 100 = -12.12$

(ii) $3.7 - 1.8 = 1.9 \text{ m}$

e The student calculates the percentage and the difference in metres correctly and shows their working. **4 marks**

(b) The figures show data that have been collected for both the extent and the thickness of Arctic sea ice. When using the data, people have to be aware that there may be factors that affect data reliability and validity. The Arctic Sea is a very large area and so sea ice was <u>difficult to measure accurately</u> before satellites were used. Submarines were used to make observations (e.g. between 1958 and 1976) but the extent of their measurements depended on <u>where they could sail</u>, e.g. US submarines during the Cold War. Today's <u>satellites</u> can make many observations each day throughout the year and so produce more detailed information. Thus the reliability can be <u>different within a data set</u>. However, different groups may use <u>varying methodologies</u> to interpret satellite imagery for data analysis and this might affect the reliability of the data presented in the figures.

e The student identifies issues that may have affected the consistency of the data collection and interpretation, and uses an example from the resources to illustrate the point. **4 marks**

'difficult to measure accurately' Data collection difficulties.

'where they could sail' Expansion of point linked to data collection.

'satellites' Links to developing technology.

'different within a data set' Links to data reliability.

'varying methodologies' Links to differing methodologies of interpretation of raw data sets.

Student B

(a) (i) $\dfrac{14.5 - 16.5}{16.5} \times 100 = -12.12$

   (ii) 1.6 m

ⓔ **The student calculates part (i) correctly. Part (ii) is incorrect and the student does not show their working.** **2 marks**

(b) The data in Figure 5 and Figure 6 may not be reliable because different instruments may have been used to collect the same type of data in various places. The equipment may have been <u>set up differently and used by a range of people</u> who might take readings in dissimilar ways. This might have caused slight variations in the accuracy of the data. These variations may not have been accounted for in the final readings.

ⓔ **The student shows that they understand why data sets may be unreliable. They focus on the set up of the data collection equipment and also mention chances of human error. However, the answer is not clearly related to sea ice data collection. The student could mention the limitations of collecting data by submarine and/or satellite (Figure 6).** **2 marks**

   'set up differently and used by a range of people' Understanding of how differences in the use of equipment may affect data quality.

Question 3 mark scheme

8 marks (AO1 = 4 marks, AO3 = 4 marks)

This question asks you to look at selected data from four countries and to analyse differences between their oil and gas reserves. Suggested ideas are outlined below, but you do not need to include all of these in your response. Other relevant points will be given credit.

AO1 Demonstrating your knowledge and understanding
➤ Countries have different levels of oil and gas reserves as a result of factors such as resource distribution and technology.
➤ Levels of oil and gas reserves can change over time.
➤ The reserves-to-production ratio (R/P) indicates the future availability of a resource depending on levels of production.
➤ Changes in the amount of proved reserves can be linked to exploration and production rates.

AO3 Using relevant skills to analyse data
➤ The data present a mixed view of oil and gas reserves for the selected countries.
➤ Canada has the largest amount of proved oil reserves and the highest R/P ratio. Its known reserves increased significantly between 1995 and 2005. Russia also has high levels of oil reserves but these have reduced slightly since 1995.
➤ Russia has by far the highest proved natural gas reserves and this has increased slightly since 1995. It also has the highest R/P ratio. Although Norway has the lowest amount of proved gas reserves, its R/P ratio is higher than that of the USA and Canada.

Answers to this question will be given a mark within a level band

Level 1 (1–3 marks) You show isolated knowledge and understanding of the differences between oil and gas reserves for the selected countries, with some inaccuracies. You make a limited attempt to analyse the data and make few geographical connections.

Level 2 (4–6 marks) You show some knowledge and understanding of the differences between oil and gas reserves for the selected countries, possibly with some inaccuracies. You make an attempt to analyse the data and make mostly relevant geographical connections.

Level 3 (7–8 marks) You show accurate and relevant knowledge and understanding of the differences between oil and gas reserves for the selected countries throughout. You analyse the data thoroughly and make relevant geographical connections.

Hints and tips

Which countries have the most and fewest reserves? How do these reserves change over time? How does the R/P differ between countries?

Question 3 example responses

Student A

The four countries shown in the table have different amounts of proved reserves for both oil and gas. Proved reserves include the amount of the resource that is known to exist and which it is economically viable to extract. Different countries have varying amounts of natural resources. This is due to factors such as the underlying geology and the extent of the territory of the country.

In terms of oil reserves, Canada has by far the largest quantity (172 thousand million barrels in 2015). This is over three times the amount it had in 1995. Its R/P ratio is also far higher than the other selected countries, indicating that under current conditions, Canada's reserves will last significantly longer than those of the others. The USA and Norway have significantly different amounts of oil reserves in 2015 (55 thousand million barrels and 8 thousand million barrels, respectively) but have similar R/P ratios. This could be due to the USA having a higher rate of production than Norway. Norway's reserves have declined slightly since 1995, so the government may have decided to slow down production to maintain its reserves.

On the other hand, Russia has the largest amount of natural gas reserves in 2015 (32.3 trillion cubic metres) and this has remained relatively stable since 1995. Canada and Norway have the lowest amounts in 2015 (2 trillion cubic metres and 1.9 trillion cubic metres, respectively), with reserves fluctuating slightly since 1995. The USA has seen a rise in natural gas reserves since 1995 from 4.7 trillion cubic metres to 10.4 trillion cubic metres. However, its R/P is slightly lower than Norway, indicating that it has higher rates of production.

However, better technology may enable countries to undertake resource exploration so that they can maintain or increase their reserves. This may change the R/P ratio in the future.

ⓔ The student analyses the data effectively and attempts to analyse the differences in data between the selected countries. They show good understanding and make connections between the data and their own knowledge. Examples from the table are used to show changes over time for particular countries as well as comparing selected R/P between countries. General reasons why the figures may vary between countries are suggested. **Level 3, 8 marks**

Student B

The selected countries all have interests in the Arctic region but their reserves of oil and gas include resources found across all of their land and sea area. Russia is the largest country in the world and has large reserves of non-renewable resources. The table shows that Russia has the second highest amount of oil reserves after Canada (102.4 thousand million barrels) although this has gone down since 1995, whereas Canada's oil reserves have increased a lot. Canada has the highest amount of oil reserves. The USA has almost doubled its reserves since 1995. This could be because it is spending a lot of money on new technology to find new areas where oil can be developed. Norway has the lowest amount of oil reserves. Maybe this is because it is a much smaller country than the others.

Russia has the highest amount of natural gas reserves (32.3 trillion cubic metres) followed by the USA (10.4 trillion cubic metres). Whereas Russia has not increased its reserves by very much, the USA has doubled its reserves since 1995. Again, the USA could have invested money in exploration. Canada and Norway have much lower levels of natural gas reserves.

ⓔ **The student describes differences in total proved reserves in oil and gas between the selected countries. They make connections between the data and their own knowledge, and put forward some logical ideas. Relevant data are used to justify points made. The student does not comment on R/P values for either oil or natural gas. This would show more thorough analysis and lift the answer into Level 3.** **Level 2, 5 marks**

Question 4 mark scheme

8 marks (AO1 = 4 marks, AO3 = 4 marks)

This question asks you to analyse the data on military and health expenditure for selected countries. There are significant differences between the expenditure on military and health of the countries shown. You should show some understanding of the graph as well as suggesting reasons for the patterns shown in the graph. Suggested ideas are outlined below, but you do not need to include all of these in your response. Other relevant points will be given credit.

AO1 Demonstrating your knowledge and understanding

➤ Governments can decide what they spend their money on, e.g. health, education, public service and defence.
➤ The proportion of government expenditure can be used to help understand the priorities that different governments set for their countries.
➤ Superpowers and emerging superpowers have a high proportion of spending on military equipment and personnel as they may wish to protect their interests or expand their influence.
➤ A high proportion of healthcare expenditure may be linked to factors such as government social policies, investments in the healthcare system, high wages and cost of medical technology.

AO3 Using relevant skills to analyse data

➤ In general, the countries selected spend a significantly higher proportion of government money on health than military, with the exception of Russia, which spends more on its military. The USA only spends a slightly higher proportion on health than its military.
➤ The superpower countries (USA and Russia) spend a significantly higher proportion of government money on military expenditure than the other countries.
➤ The USA and Sweden have the highest proportion of government expenditure on health.
➤ The countries selected have significant interests in the Arctic.

➤ Data have been taken from two different years (2011, 2012) so are not directly comparable.

➤ Data may have been collected in different ways by different countries and so there may be discrepancies. Military spending is a particularly sensitive area and so there may not have been a full disclosure of the costs.

Answers to this question will be given a mark within a level band

Level 1 (1–3 marks) You show isolated knowledge and understanding of the similarities and differences in military and/or health expenditure for the selected countries, with some inaccuracies. You make a limited attempt to analyse the data and make few geographical connections.

Level 2 (4–6 marks) You show some knowledge and understanding of the similarities and differences in military and health expenditure for the selected countries, possibly with some inaccuracies. You make an attempt to analyse the data and make mostly relevant geographical connections.

Level 3 (7–8 marks) You show accurate and relevant knowledge and understanding of the similarities and differences in military and health expenditure for the selected countries throughout. You analyse the data thoroughly and make relevant geographical connections.

Hints and tips

Which countries have the highest proportion of government spending on military and health? Are there patterns shown? Can you suggest reasons for the differences between countries on the graph? Why might some countries have a higher percentage of military expenditure than others? Why might the data be unreliable?

Question 4 example responses

Student A

The graph shows the proportion of government spending on two particular areas, namely military and health. However, there are similarities and differences in countries' spending priorities in these two sectors.

In terms of health expenditure, nearly all of the governments of the countries shown spend a higher proportion of their money on healthcare than on the military (ranging from around 19% to around 12%). This might be because they have social policies which focus more on the health and welfare of their citizens (e.g. Sweden at just over 19%) or their healthcare systems are expensive to staff and to equip with the latest technology. The exception to this is Russia, one of the world's superpowers. It spends the lowest proportion on healthcare (10%) as its system may be more underdeveloped in some regions or it makes other sectors of the economy a spending priority.

The Nordic countries spend a significantly low proportion of their government expenditure on the military (under 5%). Military expenditure may be reduced if countries perceive there to be less threat to their security so they do not need to invest heavily in military hardware and manpower, or they may invest in intelligence and surveillance, which may be less expensive than traditional weapons. However, the two superpowers, the USA and Russia, have by far the highest military spending (over 17% and over 15%, respectively), showing a significant commitment to maintaining an effective military force. It should be noted that the countries shown all have land or territory within the Arctic Circle and so the USA's and Russia's proportionally high expenditure may lead to military dominance in this region.

It should be noted that there may be reliability issues with the data. First, the data have been collected in different years, making direct comparisons more difficult. Second, the figures have probably been released by the governments themselves rather than by an independent organisation. The amount a country spends on military equipment and personnel is very sensitive and so these figures might not be a true representation of the situation.

⊖ **The student understands the data well and explains the main patterns for both health and military expenditure effectively. The student groups the countries to help put forward points and highlight anomalies. Connections are made between the data and the wider issue of the military in the Arctic. There is a good use of numerical data to justify points. The validity of the data presented is questioned. The answer would benefit from a concluding sentence to summarise the ideas.** **Level 3, 7 marks**

Student B

This graph shows how much governments spend on both healthcare and the military. A government spends money on healthcare to help improve the quality of life for its people. It spends money on the military usually so it can defend its territory and its interests in case of attack. Some countries also may try to expand their territory by force. This requires a lot of military equipment, which is expensive.

You can see from the graph that each country spends a different percentage of its government expenditure on both the military and health. Military spending is highest in the USA and Russia as these countries have a strong commitment to their army, navy and air force and are often involved in military actions around the world. Other nations have a much lower proportion of military expenditure, with Iceland having less than 1%. This may be because Iceland has not had recent conflicts with other countries and so does not need to invest in its military.

All of the countries have a higher proportion of government expenditure on health than the military (except Russia). It is very high for Sweden, Norway and Canada. Russia has the lowest percentage.

⊖ **The student understands the main patterns in the graph. Care should be taken when referring to 'how much governments spend' as the graph shows proportions and not amount spent. The reasons for military and health spending should be linked more clearly to the graph. An attempt is made to use examples to help explain points. The link between high proportion of military spending and superpowers could be made more effectively. The section on healthcare shows understanding but is brief. Grouping countries with similar characteristics together would help broaden the answer.** **Level 2, 5 marks**

Question 5 mark scheme

18 marks (AO1 = 3 marks, AO2 = 9 marks, AO3 = 6 marks)

This question asks you to evaluate the impact of the Eyjafjallajökull volcanic eruption in terms of its global effects. Suggested ideas are outlined below, but you do not need to include all of these in your response. Other relevant points will be given credit.

AO1 Demonstrating your knowledge and understanding

➤ The eruption had a significant impact on air travel in the North Atlantic region.
➤ This affected both people and freight in Europe and in other parts of the world.

AO2 Applying your knowledge and understanding

➤ The eruption showed the impact that a local incident can have on regions outside of the country of origin.
➤ The effect of the ash cloud on airspace led to significant flight cancellations and delays in the North Atlantic region.
➤ Some primary (agriculture), secondary (manufacturing) and tertiary (tourism) industries were affected on both a regional and a global scale.

AO3 Using a variety of different skills to analyse data

➤ Figure 13 suggests that while goods carried by air contribute only a small amount of trade by weight, they have a high monetary value.
➤ Figure 13 suggests that African countries may have lost US$65 million in revenue from perishable products because of the effects of the eruption.
➤ Figure 14 suggests that 300 airports in more than 24 countries were closed and that 7 million people were affected as over 100,000 flights were cancelled. However, this is only a small proportion of both the global number of people carried by air transport and the number of air transport departures worldwide in 2010 (Figure 16).

Answers to this question will be given a mark within a level band

Level 1 (1–6 marks) You show limited knowledge and understanding of the global impacts of the Eyjafjallajökull volcano eruption, with some inaccuracies. You apply your knowledge with few connections and with limited support. Your conclusion, if any, is unbalanced. You make judgements that are not supported by relevant evidence. Your interpretation of the evidence does not show a clear connection with ideas from the geographical topics studied.

Level 2 (7–12 marks) You show knowledge and understanding of the global impacts of the Eyjafjallajökull volcano eruption, possibly with a few inaccuracies. You apply your knowledge with some connections and with some support. You draw a conclusion but your arguments may be unbalanced. You make judgements that are sometimes not supported by relevant evidence. Your interpretation of the evidence does show some connections with ideas from the geographical topics studied.

Level 3 (13–18 marks) You show accurate and thorough knowledge and understanding of the global impacts of the Eyjafjallajökull volcano eruption. You apply your knowledge with many connections and with much support. You draw a logical, substantiated conclusion and your arguments are balanced. You make judgements that are always supported by relevant evidence. You critically interpret the evidence and show clear connections with ideas from the geographical topics studied.

Hints and tips

What evidence is there of the global impact of the volcanic eruption? Which sectors were affected?

Question 5 example responses

Student A

The eruption of the Eyjafjallajökull volcano in April 2010 had an impact not only in Iceland but also in other parts of the world and highlighted the global interdependence of many economic sectors. The eruption lasted through most of April, significantly affecting air travel in northern Europe, but its effects continued after this.

Figure 12 shows the ash plume caused by the eruption. Eyjafjallajökull is in southern Iceland and the ash cloud developed across the North Atlantic Ocean and then over mainland Europe. This hazard caused the most disruption as ash from the volcano was carried into the atmosphere and spread out. Ash consists of tiny rock particles which have been ejected by the volcano and these are particularly dangerous for aircraft as they can lead to engine failure (Figure 14). The presence of the ash cloud forced the closure of airspace, which led to significant air travel disruptions. The significance of this is underlined by the fact that it was the worst level of disruption since 9/11 (Figure 13). It was estimated that over 100,000 flights were cancelled and 7 million people were affected. People had to find alternative transport in some cases (e.g. driving across Europe) or extended accommodation. However, while the area affected by the ash cloud had a large regional impact on flights, this was a relatively small proportion of the 29.6 million flights and the 2.6 billion passengers who travelled around the world during 2010, as shown in Figure 16.

In terms of economic sectors, primary, secondary and tertiary industries were affected both locally and worldwide. Agriculture was affected in Iceland as the farm at the base of the volcano (Figure 11) would have been covered in ash, and may have been evacuated, so any animals and crops would have been affected. However, by 2016 it can be seen that the farm has recovered. Fruit and flower growers in Africa were not able to transport their goods effectively and, as these are perishable products, they would have rotted with the delays in transportation. This had a significant impact on the economy as US$65 million was lost. This is especially difficult for farmers in developing countries who are trying to compete on the world market. Global supply chains in secondary industries were also disrupted, showing how interconnected businesses are when a range of components are made in different countries. For example, Japan was not physically affected by the ash cloud and yet the Nissan plant there had to stop producing one of its cars because the sensor made in Ireland could not be transported. These types of products are often specialised and valuable and so there would be a high amount of lost revenue. The flight disruption affected the number of visitors going to Iceland, particularly from Scandinavia, and this would have had an effect on the local tourist economy. However, it can be said that tourism in other parts of the world was not significantly affected because of the ash cloud's limited extent. It may even have increased as holidaymakers looked for alternative destinations during April 2010.

The global impact of the Eyjafjallajökull volcanic event was highly significant as it caused loss of income in many types of businesses around the world, and highlighted the importance of globalisation. However, much of the world would have been unaffected directly as the flight disruptions were confined to Europe.

ⓔ **The student uses a range of evidence from Section C to develop a good evaluation of the global impacts of the eruption. They clearly understand the significance of the ash cloud hazard and explain well the impacts on different economic sectors. How this incident shows global interdependence and the effects of globalisation is also demonstrated. Evidence is used to support the view that not all areas of the world were affected and a judgement is made in the conclusion. The importance of international organisations in managing an incident that has cross-border effects (ICAO) could be mentioned.** **Level 3, 16 marks**

Student B

When the Eyjafjallajökull volcano erupted, it caused massive problems around the world. The volcano spewed a large amount of ash. Volcanic ash is carried by the wind and can affect anything travelling in the atmosphere. The eruption happened in April, which would have been near the Easter holidays so many people would have been travelling and that would have made the situation worse. The ash cloud spread in Europe and it was impossible for aircraft to fly. Many flights were cancelled and some airports were closed. Many passengers had to face flight delays and could not get to where they wanted to go. This had a big impact on tourists as their holidays would have been cancelled or they may have been stranded and unsure how to get home. It was not just people that found it difficult to travel. Pieces of Nissan cars are made in Ireland and then flown to Japan. These products are light but are also worth a lot of money. When the airports closed, the components could not get to Japan very quickly. It takes too long to ship things. When the factory ran out of sensors, it had to close down production and so did not make any Cube, Murano or Rogue models for a while.

The volcano also affected the world economy as many businesses that were not in Iceland or even in Europe lost money as they could not get their goods transported round the globe. Kenyan farmers were one group that was affected. They produce flowers which only last for a certain amount of time before going off. If the product is not perfect, it cannot be sold to the customer. A lot of these farmers are very poor and they were massively affected by losing such a large amount of money.

Icelandic farmers were also affected as they had to shut their farms down during the eruption and many of their animals would have died, unless they were put inside or taken away from danger. The tourist companies in Iceland also lost money as people did not visit.

All in all, the Icelandic volcano had a high global impact which affected many people and businesses.

ⓔ **The student shows understanding of the main concepts and tries to interpret some of the sources. More specific figures from the resource booklet would help back up points. The answer could be structured to deal more clearly with the different economic sectors involved. That some parts of the world may have been less directly affected by the ash cloud is not discussed. The conclusion is brief, and the main arguments are not drawn together to make a clear judgement.** **Level 2, 10 marks**

Question 6 mark scheme

24 marks (AO1 = 4 marks, AO2 = 12 marks, AO3 = 8 marks)

This question asks you to consider the future opportunities and challenges in the Arctic region. You should use information from all of the sections in the resource booklet. Suggested ideas are outlined below but you do not need to include all of these in your response. Other relevant points will be given credit.

AO1 Demonstrating your knowledge and understanding

➤ The Arctic region is a complex environment that faces many future opportunities and challenges.
➤ Climate change is likely to affect the landscape, and oceans bring benefits and costs for a range of different players.
➤ The management of the Arctic region involves a range of players who may have both similar and different goals.

AO2 Applying your knowledge and understanding

Some possible arguments for future opportunities in the Arctic:

➤ Melting of sea ice could allow more transportation routes between Europe and Asia.
➤ Increase in annual temperatures could open more areas up for development, including exploration of resources, and lead to more employment opportunities.
➤ New opportunities for intergovernmental cooperation between Arctic nations.

Some possible arguments for future challenges in the Arctic:

➤ Threats to way of life for indigenous people.
➤ Disagreements about future developments may cause political tensions.
➤ Negative environmental impacts from resource exploration.
➤ Increased temperatures may lead to changes in landscape, including permafrost extent, and ecosystems.

AO3 Using a variety of different skills to analyse data

➤ Figure 7 suggests a link between the decrease of September sea-ice area and an increase in carbon dioxide emissions.
➤ Figure 8 shows higher levels of global temperature anomalies in the Arctic region than in other parts of the world.
➤ Figure 9 shows a decrease in the projected summer sea-ice and permafrost extents.
➤ Section D: Various indigenous peoples live in the Arctic. Climate change could lead to developments in minerals, transportation and tourism, bringing both opportunities and challenges. The Arctic is strategically important and international cooperation is promoted by the Arctic Council.

Answers to this question will be given a mark within a level band

Level 1 (1–6 marks) You show limited knowledge and understanding of the Arctic region and the opportunities and challenges it faces, with some inaccuracies. You apply your knowledge with few connections and with limited support. Your conclusion, if any, is unbalanced. You make judgements that are not supported by relevant evidence. Your interpretation of the evidence does not show a clear connection with ideas from the geographical topics studied.

Level 2 (7–12 marks) You show knowledge and understanding of the Arctic region and the opportunities and challenges it faces, possibly with a few inaccuracies. You apply your knowledge with some connections and with some support. You draw a conclusion but your arguments may be unbalanced. You make judgements that are sometimes supported by relevant evidence. Your interpretation of the evidence does show some connections with ideas from the geographical topics studied.

Level 3 (13–18 marks) You show mostly accurate and relevant knowledge and understanding of the Arctic region and the opportunities and challenges it faces. You apply your knowledge with some connections and with some support. You draw a conclusion that is supported by an argument which is sometimes unbalanced. You make mostly valid judgements that are mostly supported by relevant evidence. You critically interpret the evidence and show some clear connections with ideas from the geographical topics studied.

Level 4 (19–24 marks) You show accurate and thorough knowledge and understanding of the Arctic region and the opportunities and challenges it faces. You apply your knowledge with many connections and with much support. You draw a logical, substantiated conclusion and your arguments are balanced. You make judgements that are always supported by relevant evidence. You critically interpret the evidence and show clear connections with ideas from the geographical topics studied.

Hints and tips

Select appropriate resources. Work out what future opportunities and challenges the Arctic might face. Are these social, economic, political, environmental?

Question 6 example responses

Student A

The Arctic is a complex environmental and political region. Located in the extreme north of our planet, the region is susceptible to very low temperatures (e.g. −30°C in winter, rising to around 7°C in summer in northern Canada). Permafrost and sea ice are present in large extents because of these low temperatures, and these conditions have presented a significant challenge to development in the past. However, changes in climate have caused environmental changes on the surface. The extent of sea ice has decreased over the past 35 years because of increased melting linked to increases in global temperatures. The extent of permafrost has also decreased. For example, parts of southern Greenland are likely to experience tree growth and lack of permafrost in the future. These changes in climate offer a range of economic opportunities for development in the Arctic but also pose significant social and political challenges to a region that includes eight countries and many different groups of indigenous people.

Increased melting of the sea ice, as shown in Figure 9, is likely to offer a range of economic opportunities to the region. Transportation routes across the Arctic Ocean are likely to open up as the presence and thickness of the sea ice decreases, particularly in the summer months. This offers the potential for faster, safer trade routes from Europe to Asia, particularly if ships' ice-breaking technology improves. The vast potential for financial gains for both governments and TNCs from unexploited oil and gas could open up new energy sources for Arctic countries and make them less dependent on sources from the Middle East. It could also provide export opportunities, e.g. Russian pipelines to Europe. Figure 19 suggests that up to 30% of the world's unexploited gas reserves are found in the Arctic. Tourism opportunities may continue to develop, particularly in areas suitable for cruises. This will bring economic revenue and employment potential to more remote communities and local tourism operators. Changes in climate may also extend the visitor season.

Future mineral exploitation would continue to be challenging in such a cold climate. Some may argue that the resources are 'fossil fuels' and are likely to increase CO_2 emissions when used, leading to more global climatic problems. There may be disputes over who has the right to exploit the resources and more data on the location of the reserves would be needed to examine this issue. Permafrost changes may also make development of energy infrastructure more difficult as the stability of the ground may decrease, making repairs more expensive. Parts of the Arctic are also susceptible to tectonic activity, e.g. Iceland, so future developments in these areas would have to be prepared to withstand the effects of volcanic eruptions and earthquakes. On a social level, future projects should take into consideration the needs of the indigenous people. Figure 18 suggests that they have been negatively affected by development driven from outside in the past. Future decisions should be made following consultations through organisations such as the Inuit Circumpolar Council, which now allow for improved representation of these people on an international stage. However, 'who has the right to do what in the Arctic' is disputed and this is one of the challenges affecting the future direction of the Arctic. The eight different countries may have different territorial claims depending on their coastlines and historical interests. They may have different views as to how the Arctic should be managed and what, if any, development should take place. The indigenous people, who have lived on the land for centuries, also have an interest in future plans. Some countries, such as the USA and Russia, have large military expenditures. While the figures given in Figure 17 would probably include spending to protect all regions, the strategic importance of the Arctic could increase the likelihood of additional military investment in the region by some countries. It is important that intergovernmental organisations such as the Arctic Council provide a forum for discussion and cooperation. Figure 20 puts forward a range of priorities from the members of the Council. There are common themes here including working with indigenous people, development sensitive to the region's environment and the need for peaceful cooperation between different parties. The success of the Arctic Council is likely to have a significant bearing on the future direction of the region, as sustainable management of the region will be reliant on effective working partnerships within the region.

It can be said that the future opportunities are mainly economic but may have limited sustainability, e.g. fossil fuel exploration. The influence of a range of players including the role of governments and TNCs will affect the scale of the environmental impacts of any future development. The fact that the region includes a major superpower (the USA) and arguably an emerging one (Russia) could lead to additional political tensions in the region. The involvement of the indigenous population will also be key to successful management of the region. These social and political challenges will have to be overcome if the Arctic is to have a peaceful and sustainable future.

e **The student shows a good level of knowledge and understanding about the issues that affect the Arctic's future. The data are critically analysed and a balanced argument is given that conveys clear ideas of both opportunities and challenges for the Arctic in the future. Relevant evidence is selected from across the resources, and the data are challenged by the suggestion that a more precise location of resources should be available to enable more detailed examination. A balanced conclusion and judgement linked to the arguments is given. Environmental challenges from permafrost decrease could be included, e.g. potential release of carbon into the atmosphere.** **Level 4, 20 marks**

Student B

The Arctic is one of the coldest places on the Earth. In Arctic Canada the temperature can reach as low as −30°C in February, making it difficult to live and work in the region. There is clear evidence from the sources that the region is suffering from the effects of global warming. Higher temperatures in this cold climate have led to the melting of ice. Much of this ice is on the sea and this can affect the ecosystem and animals that live there. There is also a relationship between the decline of sea ice in September and an increase in carbon dioxide. These issues present opportunities and challenges for the Arctic region in the future.

No one owns the Arctic. One of the biggest problems is deciding who has the right to develop the Arctic. Some countries like the USA are very rich and powerful whereas others are smaller and may have less influence, e.g. Sweden. If countries disagree about what other countries are doing then this might lead to tensions between different governments and could lead to military conflict. Russia and the USA spend a lot of their money on weapons and these could be used to increase their influence in the Arctic and even expand their countries. This would be important if vital energy resources such as oil and gas were found in these areas. Figure 19 suggests that the Arctic is an important location for unexploited resources. If sea routes open up, countries might use the military to take control of the new trade routes.

The Arctic Council has eight members and they have put forward a range of issues that they feel are important to the future of the Arctic. Canada, for example, focuses on climate change, which is really important as the Arctic ecosystem will change dramatically if the planet continues to heat up. Canada also wants to keep good relations with the indigenous people so that they will be able to continue to make a living in the Arctic. It also wants any developments to be sustainable. Denmark and Russia also want a peaceful Arctic region which will be important for the people and businesses in the region. If there is no peace then the whole future of the region is at stake. There is a lot to learn about the area and so many of the countries focus on the need for scientific research, particularly linked to climate and the environment. One of the major challenges will be for all these countries to cooperate effectively on a range of issues. They may have different motivations and different priorities. The decisions that one government may make may have an impact on others so it is important that they work together.

One of the problems shown in Section C is the effects of tectonic hazards. Iceland is on the Mid-Atlantic Ridge, where the North American and European plates are pulling apart on a constructive plate margin. It suffers from both earthquakes and volcanoes. Alaska also suffers earthquakes. If companies are going to invest in getting resources in these countries then they will have to pay a lot of money to develop building and equipment technologies which will be able to overcome problems caused by these hazards.

Although the Arctic faces many problems, there are some opportunities that the region may have in the future. The melting of the ice will have a huge impact on the region. If this ice is on the sea, then it will be easier for ice breakers to create new trade routes. If the ice melts on the land, it may be easier to build settlements for people. Local people may make money from working for oil and gas companies.

All in all, it can be said that there will be many challenges in the Arctic region in the future. The changing climate will have a significant impact, including opening up large areas which may not have been able to be explored easily in the past. Many countries will have to work together to agree how to move forward, including whether they should develop these resources at all. I think that these challenges will outweigh the opportunities. This is because any jobs that are created will only be for a few people and will stop when the resources run out. This means that they are not sustainable in the future.

ⓔ **Some resources are well interpreted but the points made and evidence selected do not link clearly to the question. The student attempts to apply their own knowledge, making connections with other parts of the course, but this is not always clearly linked to the question. There is occasional over-reliance on the resources to make a point. Some geographical understanding is shown and some opportunities and challenges are given; however, focus is mostly on challenges and so the answer is unbalanced. A judgement is attempted in the conclusion. Discussing opportunities for the future Arctic in more detail would improve the answer.** **Level 3, 14 marks**